THE SECRET GARDEN

DAWN

TO DUSK

IN THE

ASTONISHING

HIDDEN

WORLD

OF THE

GARDEN

S I M O N & S C H U S T E R

NEW YORK LONDON TORONTO SYDNEY TOKYO SINGAPORE

THE SECRET GARDEN

DAVID BODANIS

SIMON & SCHUSTER
Simon & Schuster Building
Rockefeller Center
1230 Avenue of the Americas
New York, New York 10020

Designed by Bonni Leon

Manufactured in the United States of America

1   3   5   7   9   10   8   6   4   2

Library of Congress Cataloging-in-Publication Data

Bodanis, David.
The secret garden : dawn to dusk in the
astonishing hidden world of the garden / David Bodanis.
p.   cm.
Includes index.
1. Soil fauna.   2. Garden ecology.   3. Garden fauna.   I. Title.
QL110.B63   1992
574.5′264—dc20   92-23901   CIP
ISBN 0-671-66353-4

TO LEN AND GENE PASSELL

CONTENTS

INTRODUCTION
A N D
ACKNOWLEDGMENTS

My mother grew up on a farm in Ohio, and as a kid I used to spend long summers there. I remember tagging along with my uncles while they did their chores: they'd give me little things to carry, or tools to hold, while they worked—being intelligent men, they knew how to keep an eager nephew out of mischief—and they'd also tell me what they were doing: what they were putting down into the soil or within the farm buildings; why this plant or crop had to go just here, next to that one, there. I listened, entranced, proud to be a part of these strong men working.

But in time I went on to university, and learned in my biology classes, and chemistry classes, that what they had told me couldn't be true. Plants are not "aware" of what the other plants beside them are doing; animals, and insects, and all the other creatures on a farm, or in a garden, have no way of secretly communicating with the other ones around them either. I didn't want to disbelieve what my uncles had taught me, the feelings they had shared with me, so I just put those childhood memories aside. Left them, not to be reopened, till a far distant day.

A decade passed, and the author of a curious book—*The Secret House*—was looking for a sequel. Something outside the home would be good, something about an ordinary lawn and garden perhaps, but what was it that I could find which would be interesting to say? Those university chemistry courses had not been especially interesting; a rehash of distantly remembered textbooks about what was going on inside "Our Friends: The Plants," did not seem entrancing. But then, talking with some friends at the Oxford botany department, I began to learn about a brand-new approach opening up in this traditional field. Something about new equipment, highly sensitive and portable gas chromatography units. New techniques in computer analysis. New concepts in mathematical ecology.

What it was all showing was that certain mysterious chemicals which had long been known to exist in plants—chemicals which had long been thought to be just "junk," unnecessary extras—were in fact playing a crucial role. The plants were actually using them to sense attacks, to build up defenses, and to communicate the warnings to other parts of the plant. Thus an oak tree which has a leaf bitten by a caterpillar will sense the attack, pump chemicals into that leaf to poison the caterpillar, and then pump defensive chemicals into adjacent leaves, to guard those against attacks, too.

I began to do more research. The new techniques were also showing strange goings-on in ordinary garden insects. There were chemicals which aphids were spraying to each other to synchronize their defenses; sometimes there were even chemicals which they inserted into the bush they were standing on, designed to be pumped along by the internal transportation system of that bush and come up at a far distant point to warn the other aphids *there* that there was trouble coming.

This, I realized, would be my book. Much of the material was too new to have entered the textbooks or any hardcover books at all, and I spent months in the Radcliffe Science Library at Oxford, using their compact-disc computer-search facilities, sorting through the new journals as they were coming in, eager to get more details, more twists, on this new field. One afternoon, hunched forward deep in that cold stone building, reading about tiny ants hunched forward deep in their cold soil nests, I looked out and saw the Oxford students on a sunny lawn, playing with a football. I accepted that my identification with my material was getting too extreme, and went out to join them. But soon I ended up sidling away from the game, to sort of kneel down—well, actually to stretch right out—on the ground, and watch an ant following one of its invisible chemical trails around a grass blade. It was time to go back in.

A few journals were especially useful, and I worked through a good many years of their issues almost in their entirety. These were: *American Naturalist,* the *Annual Review of Ecology and Systematics,* the *Annual Review of Plant Physiology,* certain issues of *Phytochemistry,* certain *Symposia of the American Chemical Society* (notably series 208, 296, 330, and 399), and, especially, the uniformly excellent *Journal of Chemical Ecology.* I'd planned to thank the editors of these journals here, but on thinking about it I suspect they don't really need it. Everyone in the field's already nice to them—who doesn't want the editor of an important journal which you might publish in and get grants for to like you? Plus they get to have their name high up on the mastheads. The people who need it more, I think, are the working scientists on the refereeing panels of those journals. They hardly ever want that job. It's more fun doing your own work. But if *somebody* didn't take the time to screen the submissions, these journals wouldn't come out. So, most heartfelt: I thank you, referees.

The other journals I used were: *Advances in Ecological Research, Advances in Plant Physiology, American Journal of Botany, Annals of the Entomological Society of America, Annual Review of Ecology, Annual Review of Entomology, Ecological Entomology, Ecology, Experientia, Journal of Animal Ecology, Journal of Applied Ecology, Journal of Economic Entomology, Journal of Insect Physiology, Nature, New Phytologist, Oecologia, Oikos,* certain of the *Proceedings of the National Academy of Science, Proceedings of the Royal Entomological Society, Recent Advances in Phytochemistry, Science, Trends in Biochemical Science,* and *Trends in Research in Ecology and Evolution.*

When it was time to rejoin human beings, I'd round the corner to the Hooke Library —Oxford's excellent undergraduate science lending library. Their selection was good, and Pat Back, Jan Lock, and Vivien McEvoy were uniformly encouraging, not just to this researcher, but to a whole stream of earnest and often anxious undergraduates. It was a pleasure to see the faces of these new students lift after a friendly conversation at the checkout desk, when they realized that Oxford was not going to be an entirely forbidding place. Additional thanks go to two postgraduate students at St. Antony's: Scott Affleck, who would help whenever I asked him nicely, and Susannah Kennedy, especially, who for long months was my literary mentor. Karen Jochelson made a scale model of the garden out of deliciously edible chocolate, to sustain the author at a particularly strenuous time. It was such a nice gesture that I realized I could marry her. Kris Dahl has been a friendly agent during the whole process, and Bob Bender, at Simon & Schuster, has been a patient and skillful editor. Brandon Broll read the whole manuscript for accuracy (though remaining faults are my own), and Michael Marten's Science Photo Library did an excellent job in planning and commissioning the needed pictures.

My uncles, of course, had been right all along. It wasn't that all folk tales about plants and animals were true. But the insights that experienced gardeners, or experienced farmers, had accumulated over time were now, with these new chemical analysis devices, finally being explained in scientific terms. The whole thing reminded me of what had happened with endorphins a half-generation or so ago. Runners and other athletes had long known that there was a special feeling, a "happiness," that came after strenuous exercise. Physiologists scoffed at this, until in the mid-1970s they came up with techniques sensitive enough to pick up the tiny but immensely potent quantities of the opiumlike chemicals—the endorphins—which our brain actually does produce after long exercise, and which create the euphoric sensation similar to that of natural opiate drugs.

To actually write the material up I was lucky enough to have my Gulliver-like human couple from *The Secret House* around. I'd come to like them. All I would have to do now was to bring them out from their house and put them in the garden for a while. It didn't even have to be a long while: a single summer day would do. And it most emphatically didn't have to be an unusual garden they were in. There was no need for exotic jungle plants or strange insects. I only used what would be found in any ordinary suburban lot. And that turned out to be the greatest treat:

For it's in an ordinary suburban garden that you can see—if you look closely enough —the ecology of planet Earth.

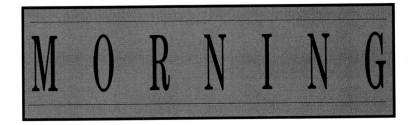

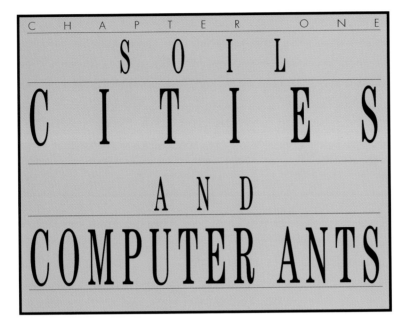

SOIL CITIES AND COMPUTER ANTS

On this early summer-day morning, the sun low over the trees, our couple are digging a small hole in their new garden at the lawn's edge to put in another geranium, in preparation for their outdoor dinner tonight. They lower the plant, carefully, slowly, the tiny naked roots dangling, into the dark moist soil, so welcoming before it . . .

Or so it seems to us.

But why should our perspective be the only one? Viewed from the new flower's roots, almost everything under the garden is cold, and dark, and not particularly welcoming. The surrounding soil is a looming cliff, and from its walls strange tunnels suddenly open with a crash. There are distant rumbling noises, as if there's some miniature city actively at work, right there, just on the other side of the underground soil barrier; there's the distinct sound of minute things burrowing closer, ready to burst through the soil to the roots at any time. Then, just seconds after the new geranium has been put in, the edge of the surrounding soil opens wider, and some crashing force from within pushes out.

Everything's so dark these six inches down that it's hard to see. There's just the blur of movement as the rumbling gets louder and closer, and then the strange E.T.-like head of a fungal creature winds out—a tired, aged miniature thing slowly turning its leathery head in the direction of this geranium newcomer. It's nearest root hairs remain still, their sensors overloaded, briefly assessing. From behind the E.T.-like fungus, deeper in the soil city again, still just moments after the new plant has been installed, there's a different sound—the blast of some gas cloud roaring down these opened-up tunnels and passageways, enough to make the fungus creature slowly bend its tired head down. For this is a poison cloud, created from another living entity deep in the soil and sent now against the newly planted intruder . . .

"Do you think it's okay down there?" the wife whispers, peering from above, from the bright sunlight down into this six-inch dark hole.

"It looks a little deep," answers the husband.

"That's what I was thinking. This plant's d-e-e-p."

He kneels a bit, beside his new-dug pit, and touches the little thing's leaves. They're soft, and oddly springy, and suddenly he's not sweaty, not uncomfortable with the dirt pushed into his fingernails, and his fingertips raw from the burlap sacks he's struggled to open, and the wheelbarrow he's had to push, but strangely happy—touched at this communion with nature.

"I think it's okay," he says.

"Well, I've got the stuff ready," she says from behind him.

"Sure?" he asks, looking at the three different sacks of soil filler.

"Oh, yeah," she says, trying not to show the wavering in her decision: taking up a small pinch of soil from beside the geranium, and holding it in her open palm as she tries to remember. "It's . . ."

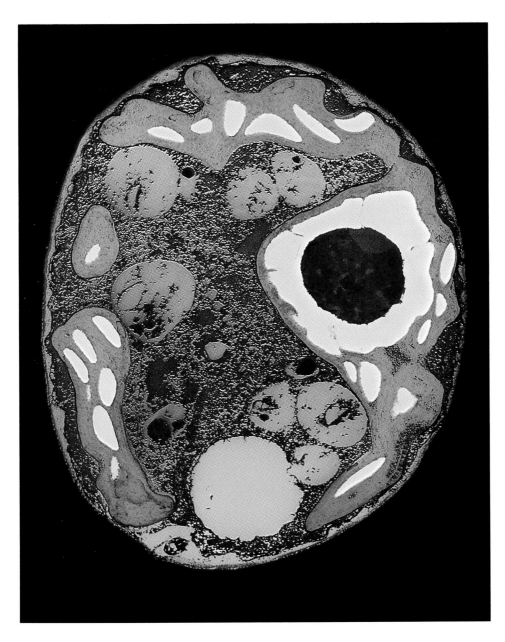

A simple algae, near-universal in garden soil. The blue and yellow circle mid-right is where starch gets pumped for storage.

The amount of life in even a tiny, half-inch-thick pinch of soil from your lawn is difficult to grasp. There will be millions of miniature insects, and 5 billion or so bacteria —about as many as there are human beings on earth. It seems impossible, since from our usual perspective there's no way so many separate living things, however tiny,

A mite. This placid creature, invisible to the naked eye, is armored for pushing through soil or dust. When the ones at the garden's surface are stirring with the dawn's light and warmth, the closely related ones inside the house (there are thousands in even the cleanest bed) are settling down to a day-long rest.

could cram into such a small space. But think of the pinch you hold in your palm as a vast apartment block. On all the ultra-narrow floor levels, there's space for microscopic creatures to live. A pinch of garden soil just a half-inch high will have space for over ten thousand such floor levels inside.

When all those floors get partitions, vertical "walls" going up to divide them into separate rooms, the total surface area contained inside a single ounce of clay comes out to about ten acres. As King David's Jerusalem was only about sixteen acres, think of the soil in your palm as a giant, space-floating city from the ancient Middle East, full of souks and passageways and vast thousands of separate, interconnecting living spaces.

The walls of this complex microcity are made up of miniature clay bricks. To keep them in place there's mortar, conveniently supplied by the gummy residues from bacteria that have the good grace to slither around a bit and then die on those many floors and walls. There's also help from the atomic forces that get amplified on this reduced scale to make each brick pulse in slow, accordionlike throbs, neatly creating electrical fields that pull in any loose, crumbling clay, so that the whole city constantly rebuilds itself.

In the largest hidden chambers are the soil mites. They look a lot like the dust mites that live at peace in our beds: both are microscopic creatures, waddling like armored personnel carriers on stumpy metal-like legs. But while the mites in our bed spend much of their time simply waiting, heads up, for the manna on which they feed (our sloughed-off skin flakes) to rain down, these mites in the soil are more adventurous. Some lumber down the dark passageways to see what other microcreatures they can catch and crunch into food. Other, less armored mites stay put, and tend gardens of leathery fungus plants, which they will later eat. They fuss and poke, pressing the chemical receptors and humidity analyzers on their heads close to see that each fungus is doing all right; jacking up the low ceilings of their chambers when things are getting too cramped; or kicking open a fresh hole in the wall, to allow in more air if the fungus needs it. Occasionally, other fungus creatures will stretch a leather arm into the mite's "garden," but this is not wise. Think of poking an arm into a hatchery tended by midget tyrannosaurs. Although these mites are slow, they can exert terrific leverage in their hinged jaws, and once they get in position, even the toughest leather arm is going to be snapped.

All these doings only worked up to full speed an hour or so after dawn, since the insulating properties of soil prevented the sun's earliest heat from reaching this depth right away. What *did* start at dawn in the garden was the bursts of oxygen spray from all the plant leaves as solar panels in their leaves started into operation. Even when the couple were upstairs asleep in their bedroom, mumbling and fussing and reflexively tugging at the sheets for extra covers, some of that oxygen was filtering in around the half-open window, and reaching their lungs. At the same time some of the carbon dioxide the couple were expelling with all their night-time air gaspings was taking the reverse course, out the window, and has by now already been incorporated into the plants. This happens so quickly that a number of the bees you're likely to see later, on waking, are loaded with sugar that includes some of the carbon you've breathed out just a few hours before, while asleep.

Spread here and there along the soil passageways are the bacteria floating on the walls. With the bacteria's small size, gravity has almost no effect, and the condensation that trickles down these walls is deep enough for them. These bacteria—close brethren of the multitudes that spend their lives swimming on our arms and face—looked like puffed-taut tiny balloons. Some have the form of little submarines with tail propellers; others look like small inflated boomerangs; yet others are shaped like ultraminute tennis balls.

And it's the gummy cement made by the bacteria that keeps all the walls in taut shape.

" . . . it's that one," she finishes, almost sure.

"Peat," he reads, accepting her knowledge. He gets ready to tip the sack over.

"No, uh," she says, really not sure. "Try that one."

"The grass clippings?"

"Y-yeah," she says.

"Sure?"

"Oh, yeah," she says. "Absolutely."

"Then here goes," he says, ready to drop them in.

This could be a Serious Mistake. Oh, grass clippings have lots of dead bacteria in them, and as we've seen, it's the gluey substance in those bacteria bodies which the soil cities use as their cement. But grass clippings should be applied months before plant-ings, and well away from any living roots. For grass clippings don't just have the glue from dead bacteria: they also have tremendous numbers of *live* bacteria on them. Those newcomers invade the microcities and battle the old populations on the tunnel walls.

The warring parties soak in nitrogen as they battle, and unless the gardener adds more nitrogen, nearby plants will suffer. The struggling bacteria also generate a lot of body heat, which produces the further problem that any plant near these grass clippings will overheat. To cool off it will start evaporating water out of its leaves more quickly than usual. In a few weeks or even days, you'll wonder why your new plants are doing so poorly, and soaking up such great quantities of water.

The peat wouldn't have been a much better choice. Admittedly, it doesn't have live bacteria to create the overheating problem, but nor does it have any amount of dead bodies of bacteria. No dead bacteria means no soil cement. All the peat can do, accordingly, is help the microbricks vacuum up extra water, and physically prop open a few of the soil passageways that would collapse from lack of cement otherwise. That's enough for some plants—azaleas, rhododendrons, and begonias will thrive on peat moss (what the British label as "moss peat")—but not for most. What the couple should have done is . . .

"Wait!" she says.

There's a sigh, deeply felt, from her husband.

Which this woman ignores, in her pride at finally, definitely, absolutely, remembering:

"That one," she says, pointing determinedly to the compost.

Thinking: He could have helped prepare for this day, too.

It's a good choice. Although compost starts out with tremendous numbers of live bacteria in it, a transformation occurs during the long months when it simmers away in a pile *becoming* compost. That simmering occurs because the bacteria inside whir through their lives and then die. The water which those bacteria had contained largely evaporates, but the gummy stuff which had held their bodies taut doesn't. It remains, and that's the main ingredient in compost. In the soil cities that gum is accordioned into place, to shore up the "brick" walls and keep the passageways open.

As the compost is mixed in, the wife drops the first soil city from her hand back to the earth, back to the hole around the transplanted geranium and roots from whence it came. Perhaps it's best that no one can see what happens now. The miniature soil city twists; it tumbles and rolls. It's a hollowed-out inhabited meteorite crashing into the earth. Clay walls collapse and crumble; sirocco winds now roar through the micro-streets and chambers; creatures get flung upside down, and, once everything finally lands with a *whap*, the impact sends clouds of whatever was on the outside back up. You can sometimes actually see this as a fine haze.

A soil fungus, magnified 400 times. The blue tubes are the feeding arms.

In this haze, like populations of suited-up astronauts hurtling away into space, there will be great numbers of fungus capsules, hardened against air buffeting and containing the DNA information to produce, when they land in the right place, another, younger, version of that leather E.T.-like fungal creature. There will also be dormant bacteria creatures. Even some of the armored soil mites will have managed to curl tight as they fell, legs buckling, going into suspended animation—a state in which they can survive for hours or days. All these spin out, flying upwards in the gusts from the impact.

Any human standing near is likely to breathe in several thousand mites, bacteria, and spores from the mix (which is why the human sneeze reflex is such a heartening thing), but most of this will pass the couple and rise up, soon catching the garden breezes and spreading elsewhere. They will land and try life someplace else around this lawn. A few might even carry up as high as traveling jetliners—on your outer window as you travel in a jet there's likely to be a number of such microscopic garden soil creatures, totally lost.

Not that the human couple, looking with pleasure at their new geranium, would know about that.

On the rosebush, just inches away, the aphids are paying no attention to these strange bipedal giants who mutter and bend around the hole at the edge of the lawn where the new plant is. They're too comfortable, attached easily to the rose stems: each one-tenth-inch aphid placidly leaning forward, shouldering its full weight head-down to p-u-s-h its feeding tube into the stem. For the rose has a network of tunnels and pumping stations deep inside it. If you fixed an ultrasensitive microphone tightly to the rose stem you could amplify the grinding, pumping, and whirlpool suctions of its micro-tunnels, carrying sugar water and liquefied vitamins around the plant—and that's what the aphids are feeding on.

The quiet, passive creatures are neatly spaced, each about a half-body width from the others—a field of miniaturized grazing cows with antennae. Every now and then one steps slightly apart and squeezes out a new baby aphid—like spaghetti extruded from a tube—and then goes back to grazing. The babies are born wrinkly skinned. In a few minutes the infant's legs and antennae unfurl, and then the little thing adjusts its feet into a good nonwobbling wide stance, scrunches *its* shoulders down, and joins the others in feeding.

So tranquil and harmonious is this early morning rose-scape of shadow and sun that it seems unfair that it should be marked by a flicking dark shadow above. But something is flying above, surveying the garden, on the hunt for vulnerable, miniature, wrinkle-skinned herbivores.

"A ladybird?" she asks, looking at the tiny sweet creature that has suddenly landed on her hand.

"Could be," he volunteers from behind, where he's concentrating hard, treading down on the soil surface to compress their new plant in place.

"Hey, definitely," she answers, pleased at the cute little thing. "There's two spots," she continues, "and there's—oh!"

He turns, but his wife is standing empty-handed, pointing to somewhere into the bushes, where the little thing has abruptly flown.

The ladybird ("ladybug," in many parts of the United States) that seemed so sweet is roaring and buzzing, its chainsaw wing motors hurtling it through the garden air. The clanking, flying death-machine switches on its ground analyzers. It was a terrible mistake to have landed on that big fleshy human thing. The ladybird's energy supplies aren't sufficient to allow many more errors like that. The flyer roars on, wobbling slightly, despite powerful hulked muscles propelling it. The wobbling is not a design mistake, but a way of getting a good trigonometric fix on the object that its analyzers have selected as a better landing surface. The killing *thing* levels quickly now, an on-board horizon indicator pulsing out signals to adjust its wing motors so that it can slow quickly enough, before it whizzes past the selected target below.

Crashing down onto the exposed side-leaf, the heavy flying vehicle lands, bent legs taking the shock of the impact, sturdy curved wing covers lifting up, under which the flying wings immediately retract. The targets, it reflexively knows, are likely to be higher up, on the freshest leaves, the ones just starting to grow, closest to the sun. The machine spreads a readying fluid over its slicing razor mouth-parts, and then this ladybird rumbles along on its two sets of tripod-tilting legs, up this rose leaf, toward its breakfast.

Since the aphids cluster together in little sheltering family herds, the ladybird machine starts with one on the outside of a circle. It clumps forward, opens wide its viciously sharp mouth-parts, and *chomp!* The ladybird begins with a nice juicy hind-quarter. Next, shall it be one of those really nibbly baby aphids? The ladybird has its pick: the poor simple aphids don't even move away, but only seem to shift in place slightly. Now there's *another* juicy one up ahead, further on within the waiting, clustering, cowlike aphid circle. Why not try *that* one? The ladybird is likely to step forward in its haste, leaving the remaining half of aphid carcass to fall over—a miniaturized cow tumbling without even a protesting miniaturized moo.

To promote this formidable but beneficial destruction, it's a good idea to leave some nettles or thistles growing wild somewhere at the back of your garden. They're a place where the predators of these ferocious ladybirds themselves have a tough time intruding, so ladybird numbers are likely to rise quickly there. As a single ladybird can eat five thousand aphids in its lifetime, having just a few plants can do a lot.

Soon the ladybird is glutted; its sagging gut half-filled with those tiny aphid legs and aphid feet. But it doesn't even bother to fly away now. It rests, seemingly safe, perhaps content in contemplation of the thousands of aphid meals to come. The ladybird is much larger than the aphids and has viciously sharp mouth-parts against their ineffectual little pumping tubes. Besides, it has spent over a dozen million years evolving into this advanced hunting vehicle. What could tiny aphid creatures possibly do against all that?

Think of that quiet, straw-hatted rural guy you've made fun of from your roaring Porsche. Now imagine your car broken down at the edge of town, and his walking to you, very very slowly, so that you see his face, and his glazed-empty, remorseless zombie eyes. You start to apologize, oh so very sincerely, but he advances at you, still zombie-rigid, hearing nothing. And you notice, as you're being backed against the weathered wooden walls of the town's general store, that all the other straw-hatted residents are jerkily advancing on you too, arms rigidly extended, down the town's dusty main street, as if they've been called by the first one.

Somehow as if they *knew*.

The ladybird has been an idiot. Aphids have been on our planet for an estimated 280 million years. They have survived meteors, large dinosaurs, ice ages, many thousands of distinct hunting beetles, earthquakes, volcanoes, and other disturbants far worse than a solo ladybird in that time. Some of their power comes from the fact that most of them are clones, and so it doesn't matter if a few get sacrificed to protect the group. Think of a brokerage firm's senior partners watching a few of their newest *junior* partners resign, after there's been public protest at the inflated salaries on Wall Street. But the aphids also get their power because of the way they can make the plant act— like a zombie —as they want:

They can make it work, for *them*.

Remember the aphids' p-u-s-h-e-d—in little pumping tubes? They're good for taking sap out of the rosebush, but they're also good for sending signals *in*. When that first cluster of aphids was attacked, and they shuffled sideways a little, they weren't just being wimps. Instead, they began to pump a warning chemical down through their feeding tubes into the rose. That chemical traveled through the rose's internal circulation system—conveniently, neatly, following all the internal passageways the rose had ready for it. Since aphids *always* feed with their heads down, pumping, soon almost all the aphids on the plant were linked by a chemical message.

The chemical signal sent in by the attacked group informs the other aphids to cluster in slightly tighter circles, so presenting a tighter defensive ring to any other ladybirds that might land. But a *second* chemical signal, similarly pumped in by the first aphids, now instructs the others to begin preparing poison droplets. Any ladybirds that stride up to the aphids will, once the signals have been acted on, be greeted by sudden droplets of the aphid stuff.

The droplets at first seem ridiculously ineffectual. But one portion of each small droplet quickly hardens—like model airplane cement—and begins to trap the ladybird, gluing its feet in place. Another part of each droplet evaporates, and gusts *over* the ladybird. It reaches waiting aphids that may have missed the tunnel-pumped instruction and prepares them to produce the gluey droplets, too.

It might not seem to be in the rose's best interest to let the aphids keep those inner tunnel channels open for their own communications. But this plant—it seems—has little control of its own workings now. The eerily defending aphids are like miniaturized chemical factories—also making vitamin extractors, lubricants, even potent weight-lifter–style steroids—and they squirt the whole gloop back inside the rose. If the plant responds as desired, the aphid masters slow down. If it doesn't, they simply increase the dosage that they inject.

Not only does this keep the tunnels wide open, it also forces the rose to work faster than usual. This is dangerous for the rose. Imagine that the straw-hatted bumpkin has succeeded in taking *you* over, so that you did everything at a whirring top speed in your city office, just to earn more money to bestow upon *him*. For your zombie controller that's great, but for you, facing an early heart attack, it's less attractive. The garden rose, speeded up by the controlling aphids, will pump more nitrogen and vitamins from the soil than it should, and the extra food allows more wrinkly aphid babies to be extruded out. What the aphids have done, in essence, is make the rose turn nutrients from soil water and plant sap into yet more aphids. Each parent aphid can produce a baby in one day, and those babies can themselves start extruding out their own babies, if conditions are right, a single day later. This is why aphids seem to appear, and grow, so quickly.

The results can be impressive. One female aphid, if unstopped, and with all its offspring surviving and reproducing, could produce several *million* pounds of aphids over a summer. This makes it clear why some judicious spraying of your plants with a cold water hose at the very start of an aphid infestation is such a good idea. And any plant of yours that has been infested, even a little bit, is going to need extra nitrogen fertilizer, soon, to keep from being wrecked.

Any wise creature that lands on the rose should try to escape from this awful,

A group of aphids, busy drinking from a plant. The two small tubes at the back ends produce the chemical clouds by which the aphids communicate; they also produce the quick-setting cement the aphids use for trapping ladybirds and other attackers.

zombie-controlled plant. For the hunter-ladybird that means a desperate attempt to disentangle its feet. It staggers and stumbles; it tries to hurry along an aphid-free leaf, as fast as it can, to open its air scoops, build launch speed, jump free into the air, and vanish.

Once that happens, the magnificent aphid-controlled defense is complete. The small tube-drinking creatures have succeeded. Their rose planet is freed. They alone are dominant, triumphant.

"Ugh," she says. "Oh, gross."

"Where?" he asks.

"There."

"Yeeuch," he says. "Ugh."

Finally now seeing those primitive, triumphant aphid bugs on their nice rose.

"Well, that's it," he says smartly.

"You're not . . . ?" she asks.

"Oh, I am," he answers, macho-strong now, even rolling up one plaid flannel shirtsleeve to show action is imminent. "I'm getting that pesticide."

"But what about last week?" she asks, plaintive, remembering a certain . . . excess, had the neighbors termed it? when she'd left him by himself.

"Maybe last week I used too little," he says, a macabre Jack Nicholson–style grin on his face.

"Hey, I'll go instead," she volunteers.

"You?"

"And why not?" she asks, in a tone meant to discourage patriarchal and sexist comments.

"Okay," he says.

Once she is out of sight, he relaxes and smiles—pleased at having manipulated her to do this errand-running—and settles on his back on the lawn and with his hands behind his neck, looking up at the morning's blue sky, content on his simple lawn.

Alone.

Among the many thousands of living creatures on the lawn that have been active since dawn and now hurry over to observe this man-meteorite, one of the very first is likely to be the tiny jumping spider *metaphidippus*. This creature—just big enough to see, if you have good vision—exists in surprising numbers, trundling down the dim, skyscraper-lined boulevards of our lawns.

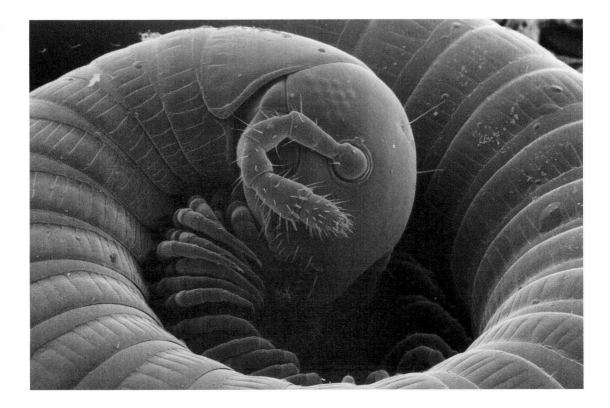

This millipede looks like it's resting, but it has actually curled into this position as a defense against attack from larger creatures on the lawn edge. Millipedes themselves are ferocious attackers of fungi, using their tough jaw attachments to bite through the feeding arms.

The creature moves like a nimble armored reconnaissance vehicle, with six search-lightlike eyes on top, constantly switching from wide-scan to powerful telephoto close-up. That vision, combined with their speed and nimble hollow legs, allows these microspiders to arrive quickly. It quickly recognizes that there's nothing *really* good to climb up on here, so rotates the top of its body away, and trundles the other way, retreating down the quiet skyscraper boulevards to join the others of its species, busy in their scanning and telephoto-observing elsewhere around the lawn.

In its place, coming pretty quickly now, are the climbing fungi. These have been struggling up the blades of grass on the lawn all morning, and when their sensitive tips detect a welcoming, warm, human scalp above, they prepare to transfer across. From their stretching tips a powerful dissolving chemical starts sizzling into a single strand of the human's hair, to provide a snug gripping point from which the rest of the fungus creature can lever itself. Once it's on, the fungus will dribble out its usual excess chemical—a powerful brain-altering alkaloid, related to LSD. In the lawn that does an excellent job of destroying potential enemies of the fungus; here, in the scalp, it will just drip over the handholds that other fungi are trying to establish, as they transfer to this man-mountain, and it will keep a few of them back. From the broken grass blades below, a few powerful steroid hormones are leaching up, too. Lawn grass makes these steroids to try to defend itself against other microscopic creatures hurrying past. The whole living horde will contentedly stay in place on your hair creature—dripping alkaloids and carving grip-holds—until the next thorough shampooing. Any broken-off bits from *metaphidippus* explorers which didn't get away in time will stick there, too.

Even before the climbing fungus is entirely established, the earthworms will likely have started tunneling their way up to join the crowd. It's not the earth-pounding thud of the landed human that alerted them, even though that *was* noticeable below, where it crashed open soil-city microstreets and chambers, flinging microscopic soil mites and wall bacteria askew. Rather, the earthworms seem more attracted to the volatile chemicals that the human is inadvertently discharging from any previous shampooing and washing and shaving. The worm warily breaks a hole through the lawn surface—there are enough around that there's likely to be one unseen, directly under the vast human —and wriggles up a naked, sensing tip. Only if light detectors on that wet tip confirm that the earthworm is safely in shadow will it rise higher.

If the recumbent human moves suddenly, say by reaching sideways to scratch a rib, the earthworm either does a quick descent, or pulses out alarm chemicals. Those chemicals reverberate down the long empty soil tunnels further below, and warn any earthworms exploring down there to shift away. It's a good solution to the problem of how to communicate long distances in a world of interconnecting soil holes. But some of the chemicals also rise *up*. That's important too: these mighty aerosol bursts can be usefully fatal to many of the earthworm's usual foes. But to the gigantic human, sniffing contendently in his half-sleep, however, they're ineffectual, registering only as part of the nice fresh soil smell we detect this close to the ground.

Where the earthworm was attracted to our body-cleansing smells, and the fungus to our warmth and hair chemicals, the next creature to arrive—the microscopic .03-inch-long tardigrade, or "water-bear"—is simply attracted by human moisture. For it is a creature that *needs* this water . . . to wake up.

Most of this little creature's existence is spent in dormancy—it's the only one of the lawn dwellers that we can trust has *not* been active since dawn. You'll find them sleeping on the drier parts of your lawn, and sleeping in the topmost soil cities, and on the moss on trees, and even, quite often, sleeping on your kitchen table where they've tumbled from the ferns and moss that comes packed with delivered flowers. Through a microscope they don't look very interesting—just something like tubby, rounded, miniaturized oak barrels. But given a supply of water they begin to change. The creature wakes, and four sets of tiny plump legs pop out, and then, like a newly woken koala bear, it searches out something comforting to cling to.

When you stretch out on your lawn in summer, or walk barefoot there, or unwrap some delivered flowers—or even just reach to the moist edge of a lemonade glass that's been left out on a low table in the garden—it's quite likely that some of these drowsy microbeasts will have found you and clambered aboard. The water-bears have moisture sensors, which are sensitive enough for them to locate, to within just a few degrees of angle, the direction from which the life-giving humidity is coming. At first they can't drag themselves very quickly, but that's understandable. It might have been months or even years that they've been sleeping before they walk over to you—a five-year sleep is common, and even one of sixty has been recorded. This dryness-induced dormancy is an excellent way for them to settle in parts of the garden which other animals have had to leave alone.

Once the mini-bears reach a safety spot such as a slightly sweaty sleeve cuff, or that wonderful waiting planet of human hair, they hug tight and start a bleary-focused assessment of the giant being they've landed on. Then there's some nuzzling of their front feeding parts up against the human to lap up the moisture they need. Then an emission from their other end, where there are *three* excretory glands to deposit their microscopic waste products—water and nitrogen, mostly—on the unseeing human. And then, once they've fed, they'll do what most living beings do and use their time on this safe living planet to search out a mate.

The drowsy *tardigrade,* or water bear. Just .03 inch long, and liable to hug onto any passing human wrist or trouser leg when woken from its rest on moss or grass. The water bear contracts into an inert barrel shape in a dry environment and can survive up to sixty years like that. With a single drop of water, it slowly revives.

There's usually some drowsy bumbling before the pudgy little males and females manage to find each other. But then the genital pores on their backs open, a fertilized egg is inserted in the proper place, and the next generation of water-bears is underway. When they're born the babies are miniature versions of their parents. They don't age as we do, but simply get larger when there's enough water. But if the parents have crawled up to an only slightly moist part of the waiting human—say that cotton shirtsleeve, instead of a good humid scalp—then any puff of air, or sudden human nostril exhalation, will send the newborns floating off, an aerial flotilla of hibernating water-bear pups, drifting, quite invisible, across the garden air over the cool green lawn below.

By now, it's almost certain that an ant will have joined the other lawn-surface investigators. But the ant stays only for a moment's sniff before hurrying off, disdainful, too busy to stay—too busy building its lawn-surface computer.

It's not something that an ant can manage on its own. An individual ant has just a few thousand brain cells, and while that's enough for a little perplexed head-turning, it's too low for building a computer. But individual ants aren't what count. Think of there being in your garden a Thing, which weighs a few pounds, and spends most of the time safely stashed away under the soil somewhere. This Thing only has the ability to send a few dozen or hundred tiny mouths out into the garden at any one time, each affixed to a small mobile carrying device. But that doesn't change the fact that the Thing, in its entirety, has several million brain cells. It's this Thing, this *assemblage* of all the ants, that does the computing.

A series of satellite pictures, if they could be sufficiently enlarged, would show what the ants have been doing. The first picture, from the earliest morning, would show no ants out then; the garden was still in cool misty shadow. But an hour later the ants would have a substantial network established through the lawn. This is what the recumbent human has disturbed, cutting their food-carrying paths smack in the middle. But let there be a satellite picture in just a few additional minutes and something most impressive will have happened: the ant network will have reconfigured itself to avoid the descended giant, and it will have done this in one of the most mathematically efficient ways possible. This is why the ants are so abundant, and in fact are generally considered one of our planet's Dominant Beings. They immensely outnumber all other noninsect terrestrial animals combined. As biologists like to put it: for every pound of us, there are several dozen pounds of *them*.

Another reason ants succeed so well is that they're superb lawn-traversing machines. When this first one backs away from the shadow of the giant human and reenters the main part of the sunny, hot lawn, little air-scoops on its side automatically switch on. A mist of cooling water vapor puffs upward from them. That keeps the ant's temperature down, but it could also mean that the ant's nitrogen—the equivalent of our urine substances—would become overconcentrated.

To avoid that, the body of the ant on your sunny lawn switches on a second series of internal pumps. These whir and concentrate the nitrogen into something like tiny crystals, which are dropped behind the ant as she walks. These nitrogen packets are like miniature battery-packs for the lawn; they soon break open, and serve as an excellent fertilizer. Only a few—probably less than two dozen—of the urinelike crystals stick right to the human's shirt, arms, and hair; most of the rest adhere only temporarily and will cascade off when he finally stands.

The third reason for the ants' success is more peculiar. The creatures seem to be walking oddly when we look down at them—all that scurrying, then stopping, with abrupt twisting of their antennae from side to side, and finally scurrying some more. But it looks odd only because we can't detect what they do. An ant doesn't see our lawn as a series of identical empty dirt-packed boulevards, with only the occasional spider or manhole-lifting earthworm somewhere ahead in the distance. Rather, on a number of those boulevards sit translucent, above-ground tunnels, stuffed with special identifying molecules. To the ant they're a 3-D path, shimmering in place. It's a solid traffic map, and the ant merely has to step into the right one. What the ant was trying to do when it stopped and tilted its antenna was to find and get back into one of those comforting self-guiding pathways.

It's a nice trick—if you're an ant—and it works because of the properties of lawn air on this microlevel. If we produced a tunnel of perfume molecules ourselves, streaming them off behind us on a city street, it would stay in place only if the air were very, very still. In anything like normal weather such a structure would quickly blow away. But not so for these ant paths: the air down at the base of the grass blade is thick enough to hold.

The traffic lines—the translucent shimmering pathways—enable ants to find food. Any ant which happens to be around here, on a random searching mission far from its nest near the house, is now in heaven. A good number of disoriented, microscopic soil mites are, almost certainly, going to be climbing up the soil not far from the human, where the geranium was just planted.

An ant. The head moves steadily from side to side to keep track of location, while the two antennae register odor molecules on the invisible above-ground paths it walks along. The creature itself is just an expendable component of the master colony, which exists as an intelligent mass, hidden under lawn or paving stones.

The ant sniff-rotates its sensory antennae a little, to test that these subvisible morsels really are crawling around somewhere in front there, then it takes a bite and chomps some soil mites. But ants aren't piggy as the ladybirds were. This first forager doesn't try to eat everything it's found so that no other ant could share the cornucopia. Instead it takes a final mouthful—that's not piggy, that's just for necessary nutrients —and then it sets out to return to the main ant nest.

But how to find it? Ants are small, and grass blades are tall. The creature needs to take a fix on the sun. Medieval sailors used an astrolabe, but that worked only when they could actually see the sun. Except for a few hours around noon, this ant can't count on doing that. Instead it has a visual filter that allows it to see the "splash" patterns of sunlight in our atmosphere, which to the human eye are quite invisible. When the garden ant tilts its head up, the sky becomes full of these patterns. Each is a directional arrow, showing where the sun really is. With just a little trigonometry and calculus—hard-wired into those few thousand brain cells—the returning ant is soon on its way.

Getting around the Landed Human takes a little time—there's some more backing up, and stops and starts—but within a minute or two, any ant which has walked up to inspect and sniff will likely have made its way around. In doing so the ant will have worked out a more convenient way back to the nest than any of the other ants know about. It would save time for the nest if the returning forager could pass on this information. But how can it do this?

The answer again lies in the pathways. Having found food, the returning ant simply lowers a nozzle from the back of its body, sometimes so hard that it scrapes down on the soil surface, and then it switches open a release valve. Fresh chemicals spray out, and—since the air is so sticky here—the released substance holds in place. It's thick enough, jelly-wobbling on the pathway, to act as a new traffic marker, holding in place behind our homeward-bound forager. In Walt Disney films a fairy would often scatter golden sparkling dust, and the dust would solidify into a real roadway, which the other Disney characters could then see and walk along. This one is real. The other ants detect the spray-thickened pathway at all the new intersections, make the correct binary decision—all they have to do is select the thickest, *new* walkway—and the computational deed is done.

That's the ant computer. It's pretty crude—there are only sixty or a hundred switches, and the whole thing is made of evaporating odor-chemicals, not purified silicon. It doesn't even have real electrons shuttling along it, and has to make do with these simple navigating ant-machines. But the several-pound ant Thing down below doesn't care. It's good enough. This network of changing pathways carries enough information to ensure the dominance these creatures have in the garden.

That's what stretches tautly around our human now, leaving him snug in its middle. That's how the separate hurrying ant-brains join with the thudding-up earthworm, the vapor-attracted climbing fungus beasts, and the tight-hugging infant water-bears in the garden's welcome for him.

It's a wondrous picture of contentment, harmony, and bliss.

Until the female human returns with the spray containers that carry labels with a picture of a green valley and warm inviting sun and the words FRIENDLY ECO-GREEN.

Not even the ant computer can know that they're containers which just a few years ago—yet with the same chemical inside—were a little more honest. ECO-EXTERMINA-TOR is what they then proclaimed.

CHAPTER TWO

PUMPING ROSES
AND
SOS LEAVES

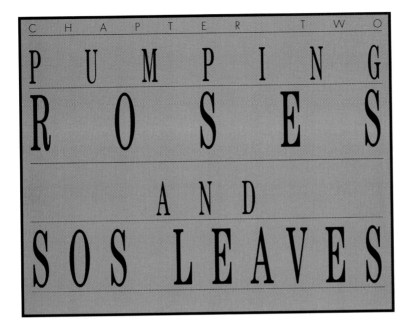

"Which one should we use?" she asks, happy, eyeing the variety of sturdy containers on the wheelbarrow she's lugged.

"Hhmnh-mmm," her stretched-flat husband mumbles.

"Hhmnh-mmm?" she repeats, perplexed.

"Hhmnh-mmm . . ." But then he wakes. "Hhh . . . hi," he corrects himself.

Sitting up on one elbow, he wonders, again, why he's had that dream of Lilliput out here.

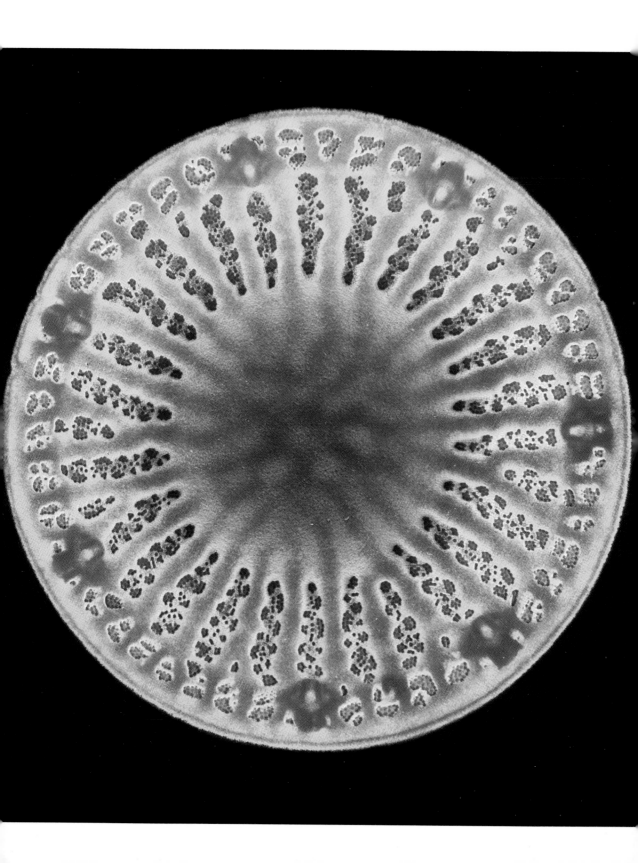

"We could use 'em all," she suggests, hopeful.

"Okay," he says. Standing now.

"Right. There's bugs . . ."

". . . on the leaves."

"So-o . . ." she begins, trying to work through the small print on the label.

"So-o . . ." he joins in, trying to read it even faster, to finish before she does . . .

"This one!" she declares.

" 'Leaf-contact insecticide?' " he reads out.

"That's it," she says, pressing the nozzle for a big arcing test spray.

Unexpected beauty. Ordinary slime on fertilizing peat is revealed as these extraordinary diatoms at over 5,000 power magnification. Diatoms are built of two silica disks, with the living body contained in the "sandwich" between. Below is the overall shape; left is a cross-section, showing the filters.

This one is totally wrong. Maybe if you're only aiming to get rid of caterpillars, and you spray this stuff at the leaves the caterpillars live and feed on, it could be the right choice. A gust of leaf-contact insecticide (look for fenitrothion as a typical main ingredient) will send sticky poison over those leaves. Caterpillars eat leaves, and so any caterpillars there would die. But aphids don't eat leaves. Their pumping tubes reach deep inside a plant's internal circulation system and only tap the nutrients there. The surface insecticide just lands on their body and rolls off.

Unfortunately for us, a lot of the stuff never even reaches their bodies. A typical pesticide applicator sends out spray particles in a range of sizes. Most will be over one hundred microns in diameter—a human hair is perhaps half that wide—and so fall almost vertically. But probably 10 to 20 percent of the liquid coming out will be in drops that are well under one hundred microns in diameter. At that size liquid spheres don't fall, but float.

Some of the floating spray will be breathed in immediately by whoever is doing the spraying. This is mildly disturbing—some of it will come out in your urine in the next day or two; much will get locked into your fat, and remain there. That alone isn't too bad. But the problem is that your own spraying isn't the end of it. What about your neighbors? Much of what *they* spray is going to end up in your body within a few days or hours, too. Then there are the farmers—it's estimated that 500 million gallons of chemical-containing mixtures float loose from crop-spreaders in the United States alone; the worldwide figure, of course, is much higher.

You can protect yourself a little bit by not spraying so far from your plants that the arcing droplets have space to scatter widely (though it's not good to spray too closely, or the plants will get scorched). In time we'll find out how medically dangerous all the stuff—which has now circled the globe and can be found in the most distant Antarctic penguins, as well as in us—turns out to be. Until then it's wise to reduce spraying with the worst chemicals, or try natural tricks such as that of reducing aphid numbers by keeping space for thistles.

For much smaller creatures, especially those tiny ones that came up to examine the humans, things are worse. The first of the very finest spray will reach the lawn surface in a steady haze over several hours. Ants are able to sniff it coming with their long antennae, and reform their path-circuitry to steer away from the stuff. But water-bears, unless they're hibernating, will probably get drenched with it, as will earth-worms, and the creatures the optical spiders eat, and all the rest. And then, days or weeks later, it drips into the soil-city tunnels—our planet's ultimate drainage system

—where those bacteria, and soil mites, and ancient fungi that supply our plants, all live. The new geranium, which is already having its problems down there from other garden residents, does not need this.

"Nope?" she asks.

"N-nope," he agrees, squinting to see. "They're still there."

"How about this one?" she suggests, lifting another of the containers.

" 'Insect-contact insecticide?' " he reads.

"They are insects," she points out, a little less sure now.

"Y-yeah," he agrees, not quite so confident either, watching her try a spray from that canister next.

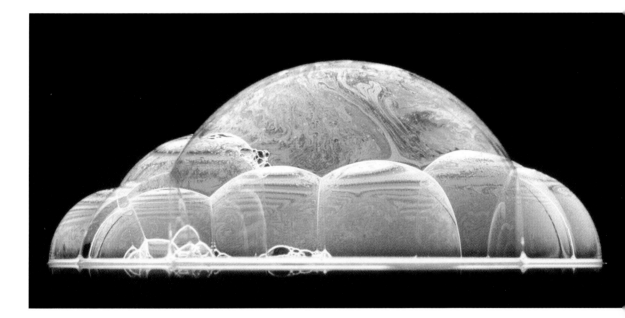

Soap bubbles. The soapy residue from pesticide applications often slips onto leaves or into soil pores; it ends up on air-dispersed droplets too. Silver colors occur where the soap layer is thinnest; green or red colors build up where the layers are thicker.

This is not a great deal better. An insect-contact insecticide will destroy any aphids it lands on, but unless you spray very very carefully, there are going to be aphids on the *under*sides of the leaves which you've missed. Because of their terrific plant-tapping communication system, the aphids will soon register that there's free space up there, on the top of their leaf world, and so, a day or two after you've sprayed and the worst of the gunk has washed off, these bottom dwellers will take over up there.

Also, what happens to any *other* creatures that have had the misfortune to be standing among the original aphids on the top of the rose leaves? They'll be destroyed by the insect-contact poison too. If it's some of the clanking ladybird machines that you hit, or the similarly aphid-eating lacewings, then there'll be even fewer of those—and especially fewer of their omnivorous young—to crunch away on the aphids. If you have to use insect-contact pesticides, the thing to do is spray as soon as aphids appear, before there are any ladybirds or lacewings around. Then you can get the top and the bottom of the leaves without damaging anything else. But you have to be careful if you're bending low to get at all the leaf parts since these sprays disperse in the air as much as any of the others.

What you'll be breathing in, with one of the most common of these pesticides, is the chemical called malathion. If you're in good health this is all right; although the chemical is toxic to human tissue, luckily we can usually break it down quickly, and almost all the constituents will be urinated out in the next day or two. But if you're not in top health, or if you've been exposed to even low doses of certain other pesticides from your neighbor's garden (especially parathion), then the systems needed to get rid of the malathion won't work properly, and there's trouble.

A slightly better insect-contact pesticide is the one called pyrethrum. It's made from a type of chrysanthemum, and so barely poisonous to us. Also . . .

"Th-they're falling!" she says.
"They're tumbling!" he says.

. . . it kills pleasingly fast. Tiny baby aphids and larger parent aphids fall over paralyzed almost immediately, pathetic rigid limbs sticking up, quickly slipping from the rose. Once on the ground, unconnected to the plant's communication system, they'll be easy game for many of the hunter insects that prowl down there. But pyrethrum also kills useful ladybirds that happen to be up there, as well as most other small living creatures.

Inside a snapdragon. With covering tissues removed shortly before the flower buds, future female parts (red), male parts (yellow), and petals (purple/green) are all clear.

If the couple had been observant, and had caught the infestation just as it was starting, none of this would have been necessary, and the old remedies of hosing with cold water, or applying soft soap with nicotine, might have been enough. But as they haven't, the treatment they should be using now is a *systemic* insecticide, such as dimethoate, or demeton methyl. You don't have to reach under the leaves with this one; you also don't have to use it only in the absence of ladybirds and lacewings. All you do is apply the stuff on a few leaves, and then it's sucked inside the rose plant. Within a week or so it spreads through the rose to all its other leaves. Aphids that poke down to the sap anywhere on the plant get dosed with it, and stagger about until they die. Ladybirds and other creatures never touch the poison directly—though they might pick up small amounts from crunching aphid legs—and so generally survive.

And sometimes, with some plants, you don't even have to do *any* of this.

"We've got them!" she cries, exulting.

"Eat this and die!" he exclaims, gloating at the creatures.

"Oh . . . wait," she says, noticing something else.

"What?"

"You see that?"

"Where?"

"There," she says, pointing to something else, deep on the shrub, just to the side of this first rose plant.

"Oh, God," he says.

"Careful," she whispers, as she circles closer.

"I know," he says, circling with her, their backs now turned to the rest of the garden, as they stare at It.

Deeper in the undergrowth, mercifully unnoticed behind the couple, there's a second rosebush: one that hasn't been sprayed, and with healthy deep roots. It's doing now what many such plants that haven't been meddled with by humans will do: it's defending *itself*—or at least trying to do that.

When the aphids first tapped into this rosebush's inner communication system, injecting their chemicals to help run things, this healthier plant wasn't really taken unawares. Rather it sensed where the sugar was leaking out, and slowed down the pumping to those parts. The aphids tried to pump faster, to make up for the diminishing flow. But the fuel for their pumping came from the nutrients that the plant provided. When a rose is fast enough, the aphids can't counterattack, because they run out of power.

The rose's next level of defense is more sophisticated. Aphids try to pump carefully, but they can't help tearing up at least a few cells when they stick their pumping nozzles in. The chemistry is still being investigated, but it seems that in many rose species, fragments of those broken cells float through the rose's internal pumping system and cause the plant to create a potent defensive chemical. It's as if a thief were breaking open a back window in your house. One of the shards of glass tumbles down into the basement, breaks open a vat of defensive chemicals waiting down there, and sets it loose so that it's soon bubbling up the stairs, to stop the thief before he can even get in.

There's plenty of this defensive chemical around in the rose, because it collects it from the tannin it makes to keep its own cells together. The name is appropriate, for that's what's been used for centuries to tan leather, in belts or boots. It's so strong that tea gets its astringent taste from tiny amounts of this tannin floating loose from the brewing leaves—it glues proteins on your tongues together. People who've sat in the sun too long get wrinkly because of something like this same tannin chemical building up in their skin.

When the rose directs its own tannin glues into the damaged area in defense, the poor aphid machines, desperately pulling away overhead, actually make things worse for themselves, pulling up more of the glue with each pump. The gushing tannin makes important proteins start seizing up within the aphids. As the aphids pump more slowly, they run out of energy, and—if the rose is really healthy—soon stop altogether.

To avoid such tannin trauma, the aphids cluster together even more than usual, to try to get so many aphid pumping derricks together, so many separate mouth-straining pulls going at once, that the rose will have to let up as many nutrients as before, despite the sticky glues. Sometimes the huddling works—which is why, in the wild, it's rare for a plant to win its battle against the aphids.

But sometimes, as in this garden now, the huddling technique doesn't work.

All about the roses, in your garden air, there are flying things which would like nothing better than a choice aphid meal. The aphids have walked into a trap: clustered together, they're now easier for these flyers to detect. And if there's enough of the ladybirds and lacewings around, that's it for most of the aphids. They'll be eaten before they can defend themselves.

But suppose this is one of those gardens where someone's been too vigorous with the poisons, and has always cleared the nettles that are home to the ladybirds. Then the aphids have the time to do something most unexpected. Look closely and you'll see them, ever so slowly, acting as if they're restless. First it'll be just one or two, right at the site of the worst clustering. But soon the warning signals, which the aphids have been injecting to each other as they feed, spread, and all the aphids get this restlessness. It's as if they're aware of what's happening, and of the crunching ladybird attack that could soon be in store for them if they stay.

The wise thing now is for the whole aphid colony, all those separate placid family herds, to retract, to try to start life again on another miniature green planet. But how? The walk to another plant is often impossibly dangerous, what with all the other creatures lurking on the garden floor below. There are a few types of aphids which actually start shrinking at this point, to try to tough it out and survive on less food. It's not a great strategy though, because a puckered, shrunken aphid is weak, and even more easily attacked. It would be a lot better if all the aphids could just fly free of this dangerous fighting-back rose. But again, how? Very few of the aphids on your rose have wings.

They need to TRANSFORM.

The adult aphids are incapable of growing wings from their stubby shoulder sockets. But if you look closer at the aphids on your rose—and again, only if it's a rose that hasn't been damaged by the wrong poisons, poured on at the wrong time, and if it's in a garden where someone's been careful to leave enough ladybirds and lacewings around to help make life untenable for them—then, as the crowding increases, you'll see the aphids stop plopping out new wingless baby aphids which look exactly like themselves. Instead, little *winged* baby aphids will start coming out.

The change is so important—the need to escape is so pressing—that even if a mother aphid is already squeezing a wingless baby down its oviduct, enough of the "crowding" chemical signal from the other aphids it's hooked up to can still transform the process. The mother aphid will produce chemicals to alter the shape of its baby, so that when the wrinkly little thing comes out and manages to stand and dry off, and unfurl its miniature antennae, it will have a set of miniature wings to unfurl with them, too.

It's hard for the new aphids to take off too early in the morning, when the air is cool and heavy. But take a look in your garden around 9 or 10 A.M., when the air is first

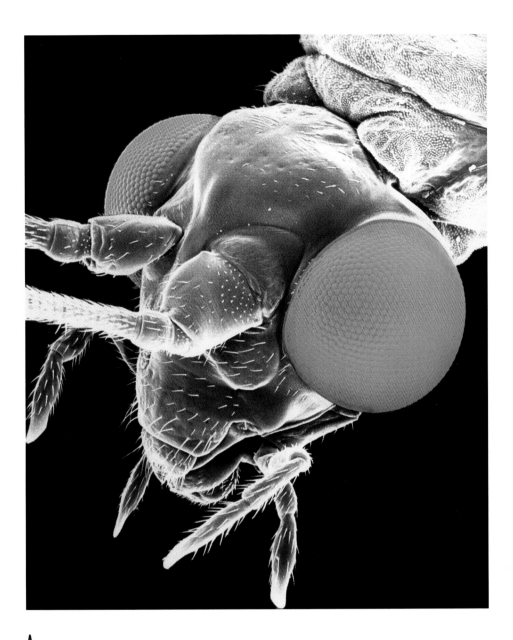

A delicate lacewing, fastidious connoisseur of aphid meals. Along with using its large eyes, it locates aphids by infrared detectors, air chemical sampling—and even the detection of rumbling movements, which the fine hairs on its head help amplify.

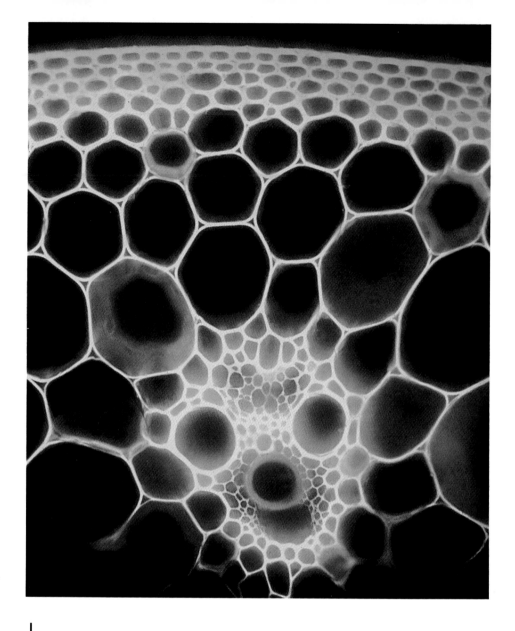

Inside a plant stem. The outermost tubes are hollow for strength. Through the inner ones defensive chemicals, sugar, and plant hormones get pumped. Photographed in fluorescent light.

getting warmer. The aphids will take their still-wet pumping tubes out from the leaf and turn to face into the direction of any wind. (A little judicious puffing then, or air-fanning with a newspaper, can help give them this slight breeze they need to get started.) With a magnifying glass you might even see the foot-pads at the base of each leg—which kept them from sinking into the waxy surface of the leaf—begin to narrow, becoming lighter and more streamlined.

Then, tiny air-scoops on their sides open wider—too small to see even with a magnifying glass. Fuel lines connect and the flight supplies begin to be fed into the wing muscles: it's sugar-water fuel, of course, freshly loaded from the rose itself. Then the creatures start flicking their wings, take a little run, maybe a jump, and then they're up in the air, tiny legs dangling, soon a vast inch or two high now over the exposed leaf edge, wing-pumping even more frantically to avoid tumbling back to the blackness of the soil looming far below. They're zooming away from this primeval rose-planet, then they're gone.

And wouldn't it be nice if other ordinary garden plants had a way of making the creatures feeding on them flee away into the air?

"Ugh," she says, getting a really good look.

"Oh, gross," he says, peering in disgust.

Seeing that primitive, gorging, webworm caterpillar on a sheltered inner leaf of their nice lawn-edge bush.

Both humans entirely oblivious of the aphid flotilla escaping from the second rose plant, behind them.

Also unaware of something else skimming, pterodactyl-like, READY to help them, in the air.

The common webworm on their bush is, like most caterpillars, a most fastidious eater. It uses antennae to sniff its food before deigning to consume anything; and then it gets neatly into position, its clean feet holding the dinner leaf steady, before it reaches down with its meticulous mouth-parts to take a single bite. There are no dangling pasta strands; no slurp-emptied soup spoons, or islets of egg yolk on jaw and moustache. If you looked with a good magnifying glass you'd see the odd, partially chewed-through fungus arm dangling out from the caterpillar's mouth; perhaps a few strings of hard yeast buds swinging in salivary globs dangling beside them. But mostly it's just a neat, flicking sharp-mouthed snip.

In some species this tidy eating has been developed to a great art. There are birds that scan a tree's canopy to find the jagged marks which a nighttime caterpillar could have left, and there's often a species of caterpillar there that makes a point of finishing off all the jagged edges from its bites, so that the leaf it has been feeding on, when scanned by a bird the next morning, will look smooth and untouched, like any ordinary smaller leaf up there on the tree.

Here on a daytime bush, the webworm is neatly clipping its way down through the top waxy surface of the leaf, through the green-glowing solar furnaces and the big oxygen caverns, into the really deep subcrust levels, where all the bush's supporting tissues and delicate tubing network and nice watery leaf juice are found. There it can feed without birds getting to it—it's protected by the outer branches of the bush, and also its taste is unappealing to birds. Exploring ladybirds won't bother it either, for although they appeared vast and clanking-dangerous to the little aphids, they're only medium-sized to the caterpillar here.

How then can a simple shrub of yours possibly get this destructive gourmet off its back? The problem is pressing. Think of all the damage you'd get from a computer-controlled excavation machine, drilling accurately through the underlayers of our city: protective concrete streets shattered open, and the nutritious electric cable tunnels below laid open for the taking. Think of the rising dust and cable fragments that would make; think, for your shrub's leaf, of the skidding solar reception cells, and broken pumping networks.

But then think also, perhaps, of the rising evaporation from all those torn leaf tissues and excavated leaf juices . . .

"I'm not going to touch it," he says, this husband-master.

"Someone's got to touch it," she says.

"You touch it."

"I'm not going to touch it."

"Well, someone's got to."

"Then you . . ." she begins—before realizing that this is getting nowhere. "You could try a stick or something."

"Hmm," he says. He kneels slightly, sighting into this bush and the waiting job, not wanting to be shown up, on their sunny lawn edge; his new domain.

"Hmmm," she says, trying to look equally sure, hoping she hasn't really hurt his feelings. Also hoping that he won't get his shirtsleeve snagged on those branches in the way.

"Hmmm," he repeats, apparently in concentration, hoping that if he waits long enough she might take over.

"What you could . . ." she starts to suggest.

But before she can finish there's a sudden rush of air, a blur—

"Huh?" both humans cry, turning to try to see. But they've missed it, and they look back to the leaf.

But the caterpillar that had been there is now somehow gone.

In the air around the garden bushes, into which the clambering aphids have escaped, there are, on a summer day, a great number of flying creatures. There might be a bee or two; the ladybirds and now aphids of course; perhaps a fruit-fly cloud. But also—in almost all gardens—there's likely to be, circling slowly, in big lazy search patterns, a much larger creature, one we can all easily see with the naked eye: *Vespula maculifrons* —the common wasp, which Americans call the yellow jacket.

On and off for hours it can circle, READY, resting only occasionally, trying this bush or that; this part of the garden or another. Its eyes are good, with down-locking radiation analyzers and wide-beam optical search apparatus. And its fuel lines are much broader than those of the small aphids, so able to pump through much more sugar-water fuel. But even so there are masses of leaves below to look through. Anything on the lower ones is likely to be blocked—to a searcher from above—by the sheltering leaves directly above it. If the wasp didn't have some additional way of picking out its targets, its prey, it would use up even its big-volume fuel supplies before succeeding.

Something nice, such as a plume of gases rising from its intended target, would be good. For the wasp already has so many adaptations for balancing and moving in the air—there are sensors for air speed and humidity concentration and the detection of potentially dangerous molecule clouds ahead—that it wouldn't be that much more difficult for it to use some of those sensors to detect such a rising gas signal from a potential target. But the ancestors of the webworm were around long millions of years ago, when identical versions of this circling wasp existed. Their webworm descendant now is certainly going to have adaptations to avoid giving any directional signal to that distant circling hunter.

The heat that the webworm produces in its chewing isn't sufficient to identify it, as that's only produced at a low level and mixes with the general heat coming up from the leaves anyway. And similarly for any bubbles of gas from the surface wax of the leaf: a leaf is always releasing microbubbles of wax on its own, so the webworm's contribution is not going to mark it out.

For the bush to survive, it will need to send up an SOS signal to call in the wasp to help. But how could it do this? Leaves are made of chlorophyll furnaces and microscopic air caverns and little tubes for carrying plant fluids. Wasps are made of electric nerve cables, and protein-rich body muscles, and radiation-analyzing brain circuitry. How could the bush make a signal, using only plant-available materials, that could float up and pass on a coherent message to the circling wasp?

It's in two steps. If a plant leaf is damaged, one of the acids that's released changes from its usual heavy form into a lighter kind which evaporates more easily. This alone would not be enough of an SOS signal to the wasp, because that acid is ubiquitous— for example, it's in lawn grass and is what gives newly mown grass its smell. Any flying creature that had evolved reflexes to swoop down to this released smell would come up with too many false alarms to keep on doing it. In a laboratory, wasps will not veer from their lazy circling if a little of this lightened acid is sprayed up.

What the wasp *will* respond to is a mixture of that smell with something else. In the leaf of our lawn-edge bush, there's another chemical mixed in. If that leaks out under normal circumstances, not much happens, because, like the acid that produces the mown grass smell, it's too heavy by itself to float upwards. And if it regularly became light enough to float it wouldn't be any good for the plant's needs either, for then it would be another too-frequent signal. But suppose it could be made in such a way that it would transform into a lighter, evaporating form *only* when it was crushed by something like the fastidious webworm caterpillar? Then the leaf would have a terrific, and immensely precise, signal.

That's what happens on your shrub's leaves at some time or another almost every summer day. When the pressure of a biting insect is applied to the second chemical, alcohols much like our ordinary drinking alcohols split loose. It doesn't matter how carefully the caterpillar bites. As long as it applies enough pressure to get into a leaf, what's inside is going to get changed. Now we know that such alcohols easily evaporate to carry an odor outward. Anyone who's had to frantically chew a strong wintergreen or peppermint mint to try to keep cops or parents from knowing that they've consumed rum or beer can testify to it. (The odors of these mints are so strong because they're carried by such modified leaf alcohols themselves.) On the leaf this telltale alcohol mixes with the mown smell, and the two then rise up, at first getting stuck in the leaf's

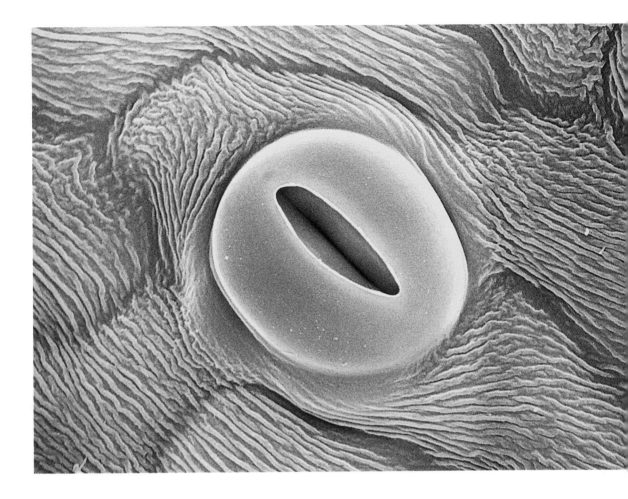

How plants breathe. These tiny pores on plant leaves open wide to pull in carbon dioxide during the day; they close when the plant is too dry. Magnification 1,200 times.

sticky microatmosphere. But then, when the webworm's damage is sufficient, more of these modified leaf juices are released, bursting free now, to slip away wherever the garden's slight breezes take them.

This explains why the waiting wasp flies in those lazy loops. Even with its excellent eyes it can't see precisely where the webworm is, hidden down there in the thousands of leaves below it. In particular, the wasp doesn't know from which bush or plant this useful SOS signal will be coming up. But once it catches a flicker of that two-chemical odor in its circling, it then does a sharp turn and prepares to follow the odor plume down. The webworm, fastidious gourmet, is slicing into the leaf without a care; not bothered by the tiny amounts of leaf juices which are trickling past its neat mouthparts and yet which are carrying the signal for its nemesis.

The wasp lowers its visored head, raises its back legs, and—only slowly at first—increases the speed of its wings. It aims toward the thicket, internal altimeter winding backwards, and then gets into the full head-down thrust of its dive. Microwinds whip past its visored head; the ground below becomes a blur as it races in. There are electrical commands cabled from its sensory units to its wings, to keep it on the right path as it hurtles through the outer leaves. There's even a bumpy compression wave of advancing air as the wasp enters the region right around the target leaf. The compression wave buffets the wasp, but its internal cables can adjust for that and keep it hurtling at maximum speed. The advancing compression wave also reaches the webworm. But the time interval is too short—there's only a few milliseconds of warning—and the webworm, even if it could begin to register what's happening, wouldn't have time to puff out any of its deterring fumes. The wasp whirs closer, vectored by the alcohols rising at the webworm's eating incision. There's a final course correction, then the wasp's jaws snap wide, and:

IMPACT!

HOOK-LIFTING!

the webworm is up, held tight, its part-excavated leaf planet quickly receding below, and only a terminal visit to the wasp nest in store.

"It could have fallen," he says, warily pushing the lowest leaves aside, looking down under the shrub now, puzzled.

"I'm sure it fell," she agrees, looking too, but only perfunctorily, eager to leave this odd, caterpillar-removing bush and head deeper into the tree-shade undergrowth, where perhaps she can help their garden some more.

The SOS signaling by the bush is so good that other garden plants use the same trick, too. When cabbages are under attack by whitefly, they release a certain gas at the point of the attack. Another sort of wasp, smaller than the yellow jacket, is often around in the garden air, and when this smaller wasp smells that gas, it vectors down and conveniently attacks the whiteflies. If you have a greenhouse it's possible to keep populations of those parasitoid wasps going, and so hold down whitefly that way. (It doesn't work as well outside though, because the small wasps die when the air temperature cools; it also doesn't work if the help-begging vegetable is weakened, as we'll see in the next chapter.)

Even trees use the SOS chemical defense. It's common to see a certain ungainly small beetle on the bark of an ordinary pine tree at some time. The creatures look so odd, almost overbalanced by their relatively big jaws, because they're not used to life out in the fresh air and sun. Normally they live within the tree, in deep dark tunnels they've bored inside the wood.

If the beetles just stayed there on the bark and waited, the tree could simply starve them out. But when the beetles crawl into their living tree burrows, they bring fungus bodies with them. The fungus gets stuck on the narrow tunnel walls, sprouts into the tree and so turns into a furry living wallpaper, and the beetle lives on that. So round one goes to the beetle.

Yet as with all the doings in our garden, what we're seeing here is not something that began recently or even just a few centuries ago, but rather something that has evolved over many millions of years. Think of every adult in a city the size of Atlanta being given a full lifetime to work on a complex puzzle, and then when each had done his or her best, passing it on—with their best partial solutions attached—for the next person to try. A group of outer-space explorers landing in Georgia at a date 60 million years in the future would be amazed at what the humans have come up with. So it is with the trees and other plants—only in this case *we're* the voyaging expedition that landed in these beings' distant future.

This means that the tree is likely to have evolved some way to fight back. To get rid of the beetles bothering it, it's been found that even a small pine tree soon begins to create poisons in the cells of its own wood. These get taken up by the fungus, the beetle feeds on the poisoned stuff . . . and round two goes to the pine.

In some species that's it. But in many common garden pines, the bark beetles change that poison to a chemical they can puff out from their hidden trunk tunnel entrance, to attract potential mates circling up in the garden air—which makes it round three to the beetle. But in these cases the pine adds an additional chemical of its own—it's a big part of the distinctive pine resins we smell. Hundreds of thousands of gallons of the sticky SOS gas are pumped out in our pine forests each year. It joins with that puff

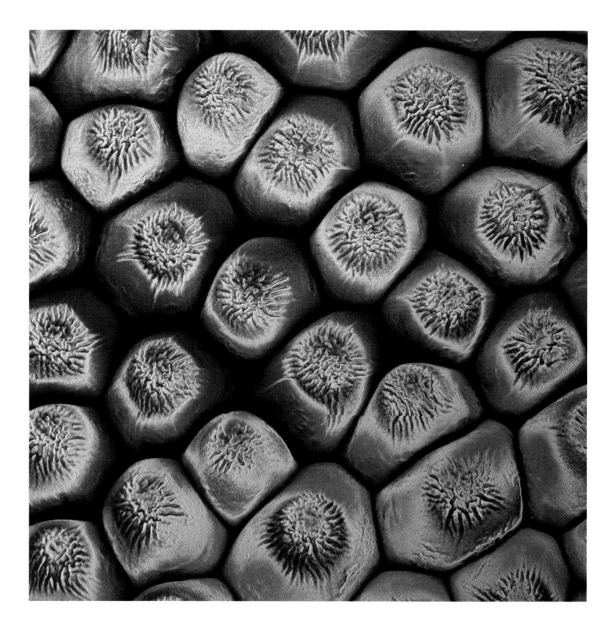

At high magnification, even a perfect rose petal no longer looks smooth. These microscopic wrinkles control the reflection of light, giving the petals their sheen.

THE SECRET GARDEN

gusting out of the beetle's tunnel entrance, so that now *predators* of the beetle can sense it and accurately vector into the tunnel to pick up these intruders . . . making it a victory in four rounds for the vapor-streaming pine.

"Is that . . ." he begins, sniffing, puzzled at the pungent resin smell which seems, somehow, to be streaming more thickly from the pine behind them now.

"Hmm?" she asks, not noticing.

"N-nothing," he says, now moving to the second, unsprayed rose. *"Do we . . . prune it here?"* he asks, knowing that any scraggly plant is just begging to be fixed.

He accidentally crushes two rough fingers into the delicate structure of a topmost bud as he asks.

"Hey, that could hurt the thing," she says, jabbing a microjagged forefinger nail into the delicate living stem tunnels to emphasize her point. *"It's . . ."*

"Here?" he asks, his massive human fingers now destructively tugging a leaf.

"No, it's um . . ." she says, really hoping she had a better memory of what she read in that gardening column in the newspaper.

"Here?" he asks.

A healthy rose pretty easily rebuilds after these minor intrusions, as we'll see later. What's more important, and can produce serious trouble, is when . . .

"I don't think you're supposed to prune it now," she blurts.

"No?" he asks, letting go of the leaf, an extra gust of invisible stress hormone rising from its damaged parts.

"N-no. We can just . . . cut one flower. For lunch."

"Oh," he says, seeing the sense of that.

Handing her the BLADE.

. . . you intentionally cut a rose during all its self-repairing, especially if you haven't washed your hands after you've been dealing with other parts of your garden. Then it's likely that something really quite small, microscopic even, will be passed from you to the open rose wound—where it will begin to grow.

At first, under the microscope, it seems to be a tiny, gnarled boulder. But then that opens, and very slowly, something emerges stretching out from within: something tired, and aging, and infinitely slow in turning its already leathery, newborn head . . .

It's our E.T.-like fungus again. Vast numbers of their capsules flew out when the soil city plummeted down during the morning replanting. When anyone walks over open soil they kick up great multitudes more, and when they touch an old branch, or the edge of a shrub, or just a garden tool that had been lying around uncleaned for a few weeks, they get even more. We live, out in our gardens, in a vast haze of these microscopic fungus capsules. There are almost always at least ten thousand of them, radiation-hardened and floating freely, in each cubic yard of the air we breathe there.

In fact, it's hard to find many points on our planet where there aren't some of the fungus capsules, slowly opening. There are unique species of fungus creatures adapted for, and currently trying out life on, almost all our flowers and lawns and trees; almost all our books and clothes and skin; our leather and timber and cotton; sometimes, even, on aircraft fuel and aircraft fuel lines and photographic emulsions.

The potential numbers of fungus capsules are hard to comprehend, for they entirely surpass what the aphids could do. One aphid, undeterred, might produce millions of pounds of aphids by the end of a summer. But that's only a single pyramid's worth. You could put it in a lucky suburb, and shine spotlights on it, and that would be that. But a *single* large fungus puffball under a garden tree, weighing a third of a pound, can contain 7 trillion spore capsules. If each one grew to its full size, there would then be a Super-Fungus, eight hundred times the volume of our *earth,* orbiting the sun where we had been.

We survive, because most die before they can grow. Virtually all of the great numbers we breathe are destroyed by protective chemicals on our tongue or throat or further inside. But if you've left dead branches on the ground from last year, so that spores get regularly popped loose from the live fungi colonizing that decaying wood, or if you're actually jamming the capsule spores from a dirty hand or clippers into a fresh cut on the rose now, then the fungus is going to have a chance to get started.

As it wobbles out from its landed spore capsule, the fungus prepares to work its way deeper into the rose. Being entirely toothless and unable to snip, it does this by producing a chemical that sizzles through the plant cells. From each capsule there's first just one extended leathery arm spraying that chemical, then soon another, and another. Since many fungi also puff out warning chemicals to each *other* from those advancing arms, the creatures even keep well spaced, for efficient team-hunting.

The curious thing is that—like us—plants almost always manage to survive these attacks. One reason is that they can often close themselves up after a cut, especially if it's on a woody stem or branch. It helps if you've given the plant a sharp angled cut: if the cut is flat rain water will stay in place, providing a nourishing pool for the fungus spores to start from.

On a leaf the plant can't do this solid closing, so it has a trip-wire system instead. If there's a fungus attack, selected cells quickly harden and die in a circle around the newly landed intruder. It's surrounded now. Unless the stretching and implanting arm moves fast enough, it's going to starve before it gets past that tiny defensive ring. Sometimes you see the small circular rings on the leaves of your plants which are a sign that these anti-invasion battles have taken place . . . and been won.

It seems perfect, but in the case of *Sphaerotheca pannosa var. rosae*—common mildew—the ancient-landed fungal creatures seem to have a way to get around that encircling. This species advances at high speeds along the surface of a rose—so fast, and surviving as much as possible on stored supplies from their landing capsule, that they can cross almost the whole leaf surface before any of the usual defensive signals get through. When this happens, you see an unpleasant white fuzz on the leaves— that's the accumulation of advancing fungus arms. You're also likely to see dark spots deeper down on the leaf: those are places where branch lines from the sprinting fungus have been sent down, to empty the living cells there.

Yet your rose usually manages to defend itself, and throw off most of even these mildew creatures. How? Sending more reinforcing chemical messages into the leaf is difficult: the fungus moves too fast for most of them to get past in time, plus there are those blocked-off dead-cell regions through which nothing will travel. The only solution is for the plant to send a faster signal: One that skims *over* its own solid leaves.

This is the second, and most extraordinary, of all the rose's natural antifungal defenses. The substance it uses is a vapor called ethylene. It's related to the ethyl gasoline used in racing cars, and is produced by a number of different cells in the rose, especially, it seems, those near the buds at the top. The stuff floats along—neatly leapfrogging any fungus in its path—and lands on parts of the rose on the *other* side of the attacking mildew fungus. This is the ideal early-warning system. The cells there start producing a fungus-destroying chemical within an hour or so of the arrival of the gasolinelike vapor. When the racing mildew fungus arrives, those leaves are defended.

We'll meet this gasolinelike vapor several times more in the garden—it's stored and sprayed out in the most unsuspected of places. Even gently rubbing a plant stem at the point where a leaf touches the stem often helps it puff out—a partial justification for the old belief that stroking your plants can be good for their health. And although high amounts of carbon dioxide stop the ethylene gas cold, *low* quantities of fresh carbon

Not strange flowers, but actually two types of fungus. The species on the right is a type of *penicillium;* the one above is omnipresent in soil. Juvenile spores are nourished from the main bodies by living tubers (blue); when they're mature they burst loose . . . whereupon more soon begin to grow. Magnification 1,600 times.

dioxide gusting over a leaf will often help the cells there produce more of the stuff. Since the air that humans exhale is around 5 percent carbon dioxide, this means that humans bending low, speaking to their plants with concern, are actually doing a useful job, simply by the life-giving carbon dioxide spray floating over from their huge nostril openings. The rose and almost every other flower use all that help in their own attempts to throw the constantly crash-landing fungus attackers off.

It all works pretty well—so long as the plants in question have the soil they need, and the sunlight, and . . .

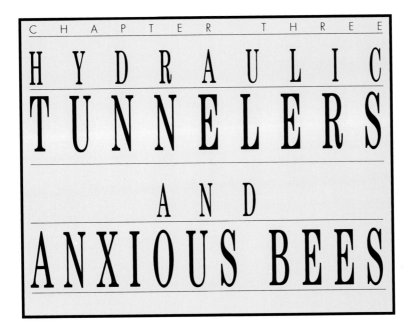

CHAPTER THREE

HYDRAULIC TUNNELERS

AND

ANXIOUS BEES

It would be great if the newly replanted geranium, standing so naked and alone, could manage to work up the same defenses as the rose and other plants. There's the same number of fungus capsules constantly crash-landing onto it from the air. And those fungi, too, are emerging from their capsules, and extending their leathery arm-probes out, and investigating, on this isolated shaken-up flower, whether conditions are right, and whether, accordingly, they can begin their quick spurt along the surface. But of course a geranium needs . . .

"That's enough," he says, this arms-folded Human Giant, standing before his geranium.
"How do you know?" she asks.
"How do I know?"

. . . water, plenty of it, to build up the ethylene that it, too, has stored among its leaves and stem. A flower this size can use a good half-gallon or more. But you still have to be careful. Pouring the water . . .

"I've read about this," she says, taking the Evian bottle from his strong hands.
She carefully, deliberately, tips it over, letting the water splash below.

. . . directly on the leaves would not be wise. For although that would splash a certain number of the new fungus creatures vast distances from their leaf-planet, it wouldn't get them all off. Instead, the water would trickle around the remaining fungi, and nourish their quickly attaching tendrils, and, in general, let them race in more quickly than they would otherwise. The thing to do is carefully, thoughtfully . . .

"Whoops," she says, giggling.
Her small hand wobbles, and the water-gush misses the leaves.

. . . to pour the water directly on the dry soil around the base of the plant instead. There it won't produce a better, moistened home for the fungi on the leaves. Instead, it will flow directly down onto the soil cities and drench the tunnels there, which the geranium roots will later reach into to pull up food. Not only will this replenish the ethylene gas supplies, but . . .

"Hey, look," he says, peering closer.
"Not . . . ?"
"Oh, no," he says, reassuring. This Male is in control.
"It's nothing," he adds, standing back.
Fairly sure he hadn't seen anything.

. . . it will also allow the flower to fight back against that fine webbing on the underside, which is the first sign of red spider mite infestation.

Another microscopic mite, here speeding through a tasty mix of soil particles, skin scales, and cat fur. The ridges on its back show its growth patterns; the nitrogen in its shell will be returned to the soil when it dies. Magnification 350 times.

The reason you can't see the mites from any distance is that they're too small. You have to get down really close and peer before you can make out the individuals: they look like stubby industrial cars, extremely small, with legs instead of wheels propelling them along. As they march, they eat into your leaves—in time turning them mottled gray, or brown—and stretch out that unpleasant fine webbing.

As those colonizing mites walk comfortably on the leaves, they're not actually experiencing the same terrain that we make out. The comfortable fuzziness which we feel on a geranium leaf is actually due to a tremendous number of tiny spikes on its surface. To us they're invisible, far too small to detect individually, but to the mites they stretch out like a huge bamboo field.

Normally that's no trouble, and the tiny mite-cars just trundle along through. What they don't realize, though, is that they're actually being *led*. Even a water-starved and manhandled geranium will still, likely, be building itself so that the exploring mites are manipulated, after a number of mazelike curves and twists, to turn up against something barring their path.

The obstacle looks like an odd little stump. If the plant is dry, and its roots haven't been fed, that microstump will be no trouble for the mites to simply step over. But if the replanting has been done properly, and the geranium is healthy enough, then there will be a colorless clear fluid, a single microdroplet, right on the top of that path-blocking stump. There will be microdroplets like that on all the other path-barring stumps, and so the other mites, scattered here and there in the leaf maze, will be facing them too.

An individual mite can't just stand there and look at the droplet. It has to get on with the business of navigating across the leaf to spread its unattractive webbing, and find good places to eat. So, poor fools, each miniature mite-car steps forward over the stump. As they do, they kick against that colorless droplet on the tip and break it open. The fluid which the geranium has so carefully pumped up now splashes out, and a dab lands on the determined mite-car.

The mite doesn't notice, at first. But this chemical which an ordinary geranium manages to produce for itself is anacardic acid. You might not be familiar with the name, but it's quite likely you know what it does. For this substance is nearly identical to the acid that builds up in much higher quantities in those plants we know as poison oak and poison ivy. Against humans, that acid causes blisters on our skin. Against the far smaller mites, massing but a fraction of a gram, stepping over and bursting open other droplet stumps as they keep on walking, it's soon fatal.

Though of course even with enough water . . .

"I can be pretty silly," she says, deciding she might as well act up to him.
Playfully, bashfully, she kicks at the ground around their first transplant patient.
As he had, so powerfully, when they first put it in.

. . . you still need to be sure a new geranium gets *access* to it. That kicking, as with any other foot-stomping or soil-crushing, is not a good thing to do. Roots look like tough woody things when you dig them up. But the main woody bit is usually just a stabilizing anchor, and not what really stretches through the soil and does the drinking. That's done, instead, by slender, microscopic, soft probes—born afresh in their thousands every day under your plants. They're so small, and the hidden soil space is so big—all those microtunnels and passageways—that their total length is extraordinary. For a single young rye plant the sum of all the root probes will be over six thousand miles, coiled there in the soil cities. The total length of all the microscopic root probes from the plants in an average garden would easily form a figure long enough to loop around the moon, and form a great figure eight in the sky. Splice together all the root probes from every plant on our *planet* and you would get the length of a thin cable that could stretch through cold space and circle around the nearest stars.

Here below ground, sprouting from a single new geranium, great numbers of the roots will snap off if you stamp boot-heavy on the surface above. Or the tiny soil cities down below will have shattered—tunnels collapsed, walls crumbled—and the soft probes won't be able to stretch through. That's why, when you put a new flower in, you're not supposed to stamp down on the soil with your foot. At most very gentle treading is called for; even better is to get down on your knees and just do some kneading and slight pressing with your fist.

But even if you've put in enough water, and even if you've done only the most considerate of toe-careful treadings, there's still one more problem. A geranium or other flower needs some help in bringing up the soil's mineral supplies for the mite-destroy-ing fluid, and the ethylene reservoirs, and all the rest. The root probes themselves can't usually pull it free on their own and use the soil city's gliding bacteria submarines to help.

Normally it's easy. Healthy bacteria sense the chemicals bubbling from any new-formed root probe, and swim over to it. They attach themselves, like tiny supply vessels berthing at a great ship, and so quickly become minute drinking tubes for the plant's

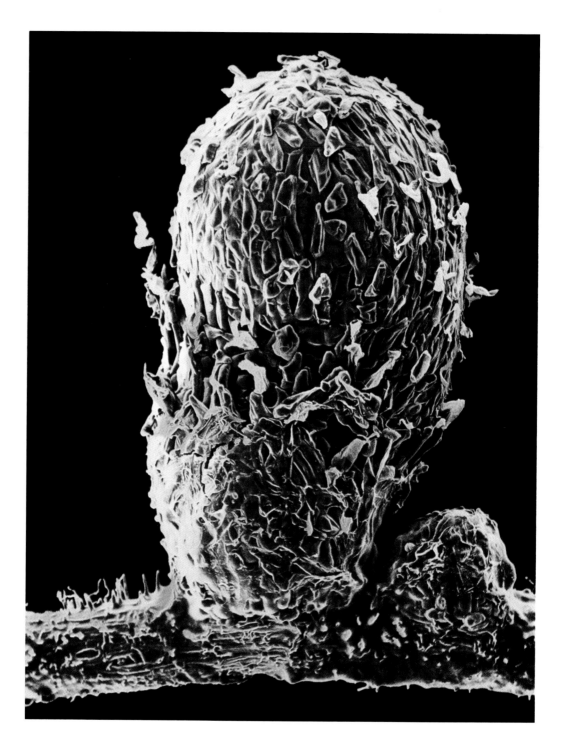

THE SECRET GARDEN

roots. Phosphates and other needed chemicals get pumped in, and the flower feeds. The problem comes, though, with the drift of pesticide droplets from the spraying earlier. The droplets might have floated in the warm zephyr air, might have hovered and danced . . . but eventually they will have landed, and rolled down into the soil.

If you've been a maniac, or even just had a neighbor who has been one, then there can be pounds of pesticide poison built up in your soil. For the bacteria and other creatures that have to wait in those microtunnels, that's terrible. The poisons burn their cranklike propellers away, or mutate them as they reproduce, or destroy them in their entirety. The struggling geranium roots will send out their chemical signals into the tunnels, but will get just an empty echo coming back.

That's why . . .

"It says something about a fertilizer here," she says, looking through the supplies back on the wheelbarrow, as he finishes the watering on this hot morning.

"Huh?"

"On the pesticide. 'This Product is Best Used in Conjunction with One of Our Instant-Power Yummy Plant Fertilizers,' " she reads.

"Something organic?" he suggests.

. . . if you've been using the pesticides regularly, you're going to have to add more fertilizers than you would otherwise, just to keep the increasingly sterile microcities of your soil supplying food. And in a case like this . . .

"It says inorganic," she answers, still reading the label.

"Hmph," he comments, knowing better, as he looks through the bags on the wheelbarrow for the superior organic in its recycled burlap sack.

. . . what you might think of as the healthiest "organic" fertilizer actually wouldn't be right. For those fertilizers need bacteria to break them down before they can work. If

Nodule on a plant root, magnified 30 times, produced by cluster of feeding bacteria. By tapping into the roots, the bacteria get to share the sugar originally created in the leaves. If the bacteria can create useful chemicals, the plant pumps out signals to lure more of them to feed.

your problem is that you don't *have* enough bacteria in the soil, that's not a wise solution. The organic fertilizer will just sit there. Doubling or tripling the dose won't help either. You've only shifted more of the expensive stuff from the fertilizer bags in your garage to the soil underneath your plants. Manufacturers like this, because they profit, but it doesn't do much for the suffering flower. And if you really pile on the fertilizer, you'll only "scorch" the plant.

The better step now for a poorly recovering flower, caused by a botched replanting, is to apply a liquid inorganic as a dressing around its base. It acts like a drip-feed for a hospital patient: everything's predigested, broken into the smallest, easily usable parts, so that the plant can take it up even without bacterial help. A nice trick is to apply it as a spray that goes directly on the leaves. That gets around the problem of damaged roots, and trades on the same easy access to the inside of a plant which the aphids used earlier. With that help—plus water and time—the geranium should recover.

As the blistering white sun passes overhead, and the morning continues, the geranium's delicate root probes continue hunting in the dark soil cities for the nutrient atoms they need. More nitrogen, more phosphates, and all the others get pulled in. To each vacuuming root probe, the nearest microcities are like that reassuring space under the seat of your battered old Ford. Sometimes you grope down there, and are lucky enough to find a needed coin or old pen. Sometimes a whole cluster of coins.

Only rarely the cold fingers of someone reaching in from the other side.

In the geranium's case though, that does happen frequently. It's the microcloud of poisonous gas, which we last saw advancing down the soil city's looming tunnels. It hasn't gone away.

Indeed, with a certain amount of outside interference, it's about to be Made Worse.

"Lots, right?" she instructs, sitting on the upturned bushel basket now, watching him put in the cucumber seeds at the edge of their new vegetable patch, close beside the new flowers.

Vitamin C. Citrus fruits and green vegetables—including cucumbers—have high levels for their own chemical defenses. Humans use it for immune defense but, unable to store it, need regular supplies. Excess quantities though are useless, getting quickly excreted in the urine.

"I don't know," he says, bringing one sweat-soaked cotton sleeve to his forehead. "Those aren't doing so well." He points to the other cucumbers, planted a month earlier, which they have been watering and fertilizing—but which are still suffering, not growing as they should.

She kneels closer and touches one of the scrawny, struggling little things, perplexed at the mystery, too.

It started well enough. A cucumber plant seems an innocent thing to us—those velvety leaves, the gradually developing simple vegetable. But, as always in the garden, such meek appearances are deceiving. Even the cucumber seed it started as, even back inside its shiny foil packet was no inert, passive thing. Rather, it would have been steadily breathing for all the long months it was waiting in the packet. Here in the soil, that new seed is a dangerous tiny factory, full of machinery and stored battery supplies, waiting to do something lethal against the nearby geranium once the right operating conditions apply.

It does all this because it needs some way to defend itself. But how? It can't defend through producing thorns or unpleasant-tasting leaves—those will only exist in the adult plant, and here the thing's still a seed. It needs something smaller, something even a tiny seed can produce.

"More water?" she suggests, mournfully, as he hefts the container to pour some more.

As soon as your cucumber seed has been put in the soil, and watered, its machinery activates certain atoms that have been packed, waiting, inside the seed. The seed's machinery constructs a gas with them, pumps it to the outer edge of the seed, and slowly begins to spray it out. This is what the poor geranium was getting blasted by at the start of the day. The stuff rolls through the soil microtunnels and gets taken up by the geranium or any other encroaching roots nearby. The gas rises up through those plants' internal circulation systems, reaches their leaves, and starts to choke them. It's bad for them, but for the seed is excellent, having helped it survive through the most vulnerable stage.

But what if the geranium planted there beside the vegetable-patch border manages to fight back? It will, often enough: the flower may have been weakened by bad handling, but won't be entirely defenseless. Even an ordinary geranium—if its leaves are still a little bit green, and its stem pumping tubes even partly in operation—will piece together, from its slowly vacuuming root probes, a counter-poison to spray back against such attackers. The poison floats out of the geranium's roots, swirls around the incoming poison cloud in the narrow soil tunnels, and ends up at the attacking cucumber seedling's roots.

That doesn't give victory to the geranium though, for the cucumber is *still* protected by the miniaturized instructions it's been carrying inside itself all the way from the seed packet. The cucumber seedlings open tiny molecular hatches in their roots. Think of steel portals in a vast missile base slowly sliding open. Most of the geranium's poison cloud simply floats along inside. Then, safely sequestered, the molecular hatches slam back closed. The cucumber mixes that poison with waiting chemicals. Much of the geranium's poison is swallowed, neutralized, and digested. Instead of being pumped up to the cucumber's leaves to ruin operations there, the poison is being converted to become *another* part of the powerful cucumber machine. In a few hours it will come out of the baby cucumber again, but this time as yet another choking gas cloud loosed against the geranium. If the geranium is too close and already weakened, enough could get back to damage it, and its defenses would have counted for nought in this case—which is why putting flowers too close to your vegetables is usually not a good idea.

If something tougher than a precarious geranium is nearby—if, for example, you've placed the cucumber in a furrow alongside threatening new tomato plants—the cucumber can still defend itself. It simply sets attackers on the threat. For along with everything else in garden soil, there are also a group of microscopic creatures called nematode worms. Through the microscope these are horrid things, with up to six lolling jaws, look like Dune-planet monsters. If the soil is at all dry, they're hibernating. But a baby cucumber, if threatened, can spray out yet another chemical which wakes up and irritates the creatures enough to make them try to move away. They slither off through the tunnels of the soil cities. As the nematodes get close to the bothersome tomato's roots, they detect the carbon dioxide which the roots give off as they grow, and use that to home in on this new food. Once there they sink their tough mouth-parts in and maul it. That weakens the tomato, so that the chemical-blasting cucumber is left in greater peace.

So it doesn't seem that survival against the other plants is the problem for the scraggly cucumbers. Nor is it . . .

. . . any difficulty with the soil maintenance here. It's true that if your hoe had been left stored in the garage, and not cleaned for weeks on end, there's going to be a lot of unwanted fungus spores on it. When that hoe gets whapped down, crashing open the soil cities, those fungus capsules come with it. They orient themselves, start to emerge from their travel capsules, and even try to attach themselves to the cucumber's roots. Since they're not the usual fungi from this particular patch of soil, they could be dangerous. But this too a cucumber seedling, rocking from the repercussions of the hoe impact, can defend itself against.

Just as did the rose, the cucumber's roots now detect the incoming fungi and begin to spray out warning ethylene bursts there below the surface. Since the ground hasn't been compressed too much by the tending couple, the gas spreads through the connecting tunnels in the soil. Other roots of the cucumber detect it, and quickly—it's a matter of hours, and sometimes just minutes—rev up their defenses in response. Those root parts become so much thicker that when the leather advancing fungus arm reaches, they're too tough for the dangerous fungus to get in. (Keeping the blade fairly clean, and trying not to hack *too* deeply when there are roots around, avoids the problem in the first place.)

For the cucumber in this dense vegetable patch there can even be an advantage in having the other plants around. Suppose it were the *tomato* roots on which the fungus spores cascaded down. Tomatoes send out the same warning ethylene gas when they're attacked. The cucumber conveniently detects that signaling—"eavesdropping"—and, if its roots are close enough, can start the necessary thickening in time, too.

The problem, of course, is that the cucumbers have been attacking *themselves*. For the chemical blasts, and the root toughenings, and the portal openings are all designed to give the baby cucumber enough space for it to grow in peace. When you pack your cucumbers too densely, everything goes wrong. The gas blasts, which should be used for keeping other plants spaced away, only damage other cucumber plant roots. Those roots have the machinery to reflexively fight back and release poison blasts on their own. The first cucumber uses its portals to swallow the incoming stuff and spray more of the original poison gas back. The second cucumber responds the same way. It's as friendly and safe as playing catch with a hand grenade that grows each time you throw it.

The result is that *none* of the individual cucumber plants get all the access to the soil supplies they need. If there's a single micropool of water in one of the underground tunnels, all the cucumber roots will have to push against each other to try to get it. More of the poison gas sprays out as they get close, and the result is that all the roots are wounded and weakened. Even if they do reach the water intact, there'll still be less —not enough—for each plant. That weakens the roots, weakens the stems, and weakens just about everything on the plant.

The plant becomes naked food, open to almost every insect marauder. Under normal conditions, for example, a young cucumber plant can disorient many of the small beetles that live on its leaves. The male beetles need to find mates at one time or another in the summer and, being short-sighted, they do so by sniffing for the distinctive chemical cloud which a female of their species produces. A healthy cucumber plant—one that's well spaced from its fellow plants—is quite capable of producing a cunning duplicate of that chemical. It sprays the stuff out all around the male beetle. The insect turns about in place, desperately trying to find the *true* location of its mate in the rising haze. The male fails to find her, or at least gets delayed, and the cucumber is less likely to get chewed by hordes of baby beetles. But the cucumber's inner machinery can produce the duplicate sex chemical only if its roots have enough space to absorb all the atoms they need. In an overcrowded furrow, they won't find them.

A plant this stressed through crowding has to conserve the limited supply of water it's getting. This means, on a summer day when the air is hot, that the breathing pores on the cucumber's velvety leaves close more than they normally would. But closed pores mean less evaporation, and less evaporation means less cooling. The plant's temperature soars up. We can't see this, because we don't have sensors to detect the infrared radiation of rising heat.

But certain other creatures flying through the air, scanning the ground far below them for a safe place to land, do see this.

"Oh, no," she says, dismayed.
"What?" he asks.
But she can't even speak. She just grabs, determinedly, for that pesticide again.

Since leaving the rosebush, the aphids have been flying and gliding in the garden for over an hour. But for creatures as small as our aphids, this aerial movement is not very easy. The air is so thick to them, their wings are so powerless in it, that they're quickly using up their fuel supplies just trying to tread air and stay in place. But they *do* have microscopic devices on the side of their heads, which are constantly aimed downward, and which they use to scan the ground. For the distinctive infrared signal that marks a green plant in heat distress.

Getting down isn't too difficult. The aphids, it seems, send out a synchronizing gas or other message so that they will descend together. To make it down to the cucumber from their great height in the air, the aphids start by simply shutting off their wing motors and falling. A bare few centimeters above the cucumber surface, they suddenly twist into the wind and switch their wings back on high to reduce their landing speed, so they won't damage the delicate landing gear of their feet.

And then they've landed.

A few of the aphids immediately try to get into position for feeding. They rear their heads high—miniature baby elephants, lifting back for trumpet blasts—and then, continuing with the momentum, let their long feeding tubes crunch down into the plant. But even though this is a cucumber in distress, feeding on it is not necessarily a wise thing to do. Aphids are under threat all the time. The cucumber itself is probably too weakened to fight back, but this could be a place where a clanking ladybird might be about to visit. Even though aphids might in time rebuff the ladybird attacks, they can't just go on *permanently* letting so many of themselves be chewed up.

Instead, you'll often see newly landed aphids walk around a little on such a leaf. What they're doing is sensing the air through detectors on their spindly antennae. They'll also taste the surface of the leaf. From the combination they can tell if the plant

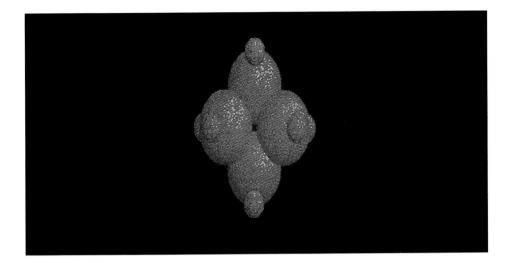

Computer graphic of the light-weight butane molecule, closely related to ethylene gas. Garden plants frequently spray ethylene from their roots when under attack by soil fungi. The vapor spreads through the soil and warns adjacent plants that an attack is imminent so they can begin preparing their own defenses. Ethylene is also the gas that spreads from one rotten apple in a barrel . . . to spoil the others.

is perhaps wounded or strained enough not to have strong defenses against them. In that case they're likely to stay. But if the couple have been sensible with the fertilizer, and put on enough water and compost—to make up for the over-dense spacing they have inflicted—then the cucumber might not be so weak. As a final check the aphids will even try to tell if a ladybird machine has been on this leaf recently. They can do this because ladybirds often leave a coded chemical on the leaf if they don't find any aphids when they first crash from the sky for their hunting. Coming back later, they'll smell that chemical, and save time by just taking off again.

The system would work fine if ladybirds were hunting for creatures that were easily manipulated. But it's the aphids, of course, who seem to do much of the manipulating. They've managed to crack the ladybird's code. If they find the signaling chemical, then the ladybird is unlikely to return right away, and so, often enough, they stay. But if they don't find it, then there's a good chance . . .

"W-whoa!" he says, restraining her arm.
"It's them," she whispers, insisting, as she grasps the plant to show him.

. . . that the aphids will lift away again, as quickly as they came, and continue their flying search till they find something better today. Rather . . .

"Where?" he demands, looking at the cucumber leaf, now suddenly aphid-free again.
"Oh," she says, trying not to show any embarrassment, and so casually looking away. Quite missing, that way, the cloud of aphids rising in the eddying heat currents behind her.
Also missing, entirely, the small whiteflies by her thumb on the bottom of that soil leaf.

. . . it's these whiteflies that take advantage of the weakened cucumber to survive. The reason they *need* to hide is that there's likely to be in your garden a population of the exceptionally small wasps that live entirely by tracking them down and feeding off them. Under normal conditions, the spikey maze on each cucumber leaf would be too tough for the whiteflies and they wouldn't even try to hide there. But on a weakened, overplanted, water-begging cucumber, something's . . . not quite right. It's not a matter of missing acid droplets, as with the geranium. Cucumbers never make those droplets. Rather, the plant is so weak—the spikes are so pliable—that the big whitefly can now safely land on this cucumber domain, and push its way in. The hunter wasp isn't used to following it there, but it is genetically programmed to follow wherever its prey goes. The whitefly stepped into the weakened maze, so it will too.

This is really not wise. The tracking wasp can follow the whitefly through almost every twist and turn in the garden air, because the wasp is much smaller and more maneuverable. Here, though, on the weakened cucumber leaf, the clumsy whitefly gets its revenge. It's strong enough to go a lot faster than the wasp. And as it marches through the maze, it's producing small gluey microdrops, which it neatly drops at intervals all along its path. The struggling wasp, following up the rear along that same path, steps right into them. First there's one droplet on a foot, then another droplet, and then—after some stumbling and lurching—this tracker is quite stuck. That leaves the whitefly free to go on feeding and glue-dribbling, and so weakening the poor, too closely planted cucumber some more.

High above, skimming forward in the blue sky, the emissary of the garden's aerial consciousness is almost at its destination. It's not aiming for the wasp or the whitefly or even that distressed human couple. Oh, the consciousness is aware of them all, and has followed their main movements out there on the sunny open ground today. But now it's the geranium it wants.

The consciousness has been keeping track of every planting in the garden in the months since the winter ended. It found today's geranium almost as soon as the couple finished putting it in. It registered the plant's location and sugar levels; it mapped the terrain immediately surrounding the geranium: the rosebush nearby and the open vegetable patch. And since then it's been calculating how to contact the geranium. Only now, a little before hottest noon, has it decided to tear off a part of its own body, and fling that out into the air, and have that soar towards the flower.

Whence our speeding emissary. But there's one thing which even the higher consciousness couldn't reckon on:

This emissary is lost.

Its first moments entering the garden's aerial environment were fine. Everything it detected below, whirring past in a blur, matched the cognitive map it had been programmed with, and also the information about where everything was supposed to be relative to the sun's angle. There was the house, and then the garden edge, and then the lawn and the oak tree, all in proper place. Even the gusting side-winds up in the air were no bother. The emissary had sensors to detect that side-shift and to compensate its flight path accordingly. It even had a sturdy belt of thin metal granules strapped on, capable of detecting the magnetic lines from the earth's molten iron core thousands of miles below even the deepest soil cities. Those invisible magnetic lines were a further course corrector.

But the problem wasn't in its on-board equipment.

"It . . . smells nicer, at least," he says, deciding to play down his powerful masculine nature, as he gestures to their dry soil terrain, its surface newly slashed asunder from the hoe.

"Oh, yeah, it's beautiful," she says, distinctly not pleased, kicking at the dumb clods of soil.

One Giant Foot impacting with the freshly exposed soil cities, releasing upward another cloud of distracting fumes from their crashed walls.

The confused emissary—that gram-sized, tubby, winged fragment of intelligence that we call a "honeybee"—slows a little. It's the representative of the *second* spread-apart consciousness in the garden, after the ants. But now what are those deep soil fumes doing up here? They weren't listed on the map the honeybee got back at the start of its flight. The Aerial Consciousness had extended parts of its own body far forward in the garden air, and had detected that the new flower was in a region where an open soil area ended beside a lawn, and from which *no* deep soil vapors were rising up.

The small flyer needs to switch on its backup recognition systems now. But that takes time, and when it was tanked-up with warm, liquid honey for this flight, it was given only as much fuel as the Consciousness—the hive, of course—had decided it needed. In this case it's only enough for a twenty-minute flight, at the bee's usual engine rate of a small fraction of one horsepower. Yet already half of that is gone, pumped steadily outward to keep the ungainly wings swimming through the thick high garden air. The honey fuel in its belly is still somewhat alive and protecting itself against bacterial infestation as it waits inside the flying bee by exploding out ultramicroscopic blasts of hydrogen peroxide. But the bee still registers its overall fuel levels steadily running down, and so must be especially alert now.

"Honey," he says.
Running one hand tenderly on his wife's neck.
Where her perfume is wafting out, reaching his nostrils.
Further masking the nectar rising up from the feebly puffing geranium beside them.

The little flyer warily banks, and descends a little. There's no sound reaching it: bees lack ears. Rather, the emissary bee responds to the signal from its odor-sensing devices. But here the signal seems blocked a little, masked by something else. The emissary waggles its front antennae powerfully to try to pull in even more of the crucial

When bees speed through a garden, this tiny creature—the braula—is likely to be clinging tight. It can't travel on its own, as over evolutionary time it has lost the use of its own wings and developed comb-like feet good only for holding onto a bee's fur.

odorant molecules. But the rising perfume is too distracting, so the bee will have to go with the *other* backup system:

It will have to go where its optical flicker-detection unit takes it.

"I know," she whispers, touched.

Then suddenly, spontaneously—and not caring who the heck sees—she flings her arms out around her terrific husband's neck, looking longingly in his eyes.

Which seem, somehow, strangely widened now.

The bee has a reading! Its flicker unit is hard-wired into its own neural circuitry to pick up sudden changes of movement below it. Any machine that cruises at high speed is going to need this. Modern aircraft have it. The bee has an extremely good ability and can detect a flicker as brief as 1/300 of a second.

In its neural circuitry, it also has maps to identify the flicker pattern that the most common plants and flowers make as it flies fast over them. But the map is never complete—many bee flights go wrong at first through poor instructions—so that the individual bee, although itself totally lacking in any overall consciousness of its task, is programmed to keep on flying around, in rough search patterns, until it reaches that intended target. The higher consciousness needs that to keep itself from becoming isolated—blind and deaf—in a constantly changing world. Admittedly a few of the mobile searchers might die in the information quest, but that's not going to bother the higher consciousness too much. Who minds losing a few skin cells from their palm when it comes to tugging open a jar of triple-chocolate cookies? And anyway, a solo honeybee might well be safe in a little exploring, especially on an object so close to where its navigation system brought it. Perhaps even this most actively flickering object below, the Large One flinging its arms forward, only a little bit away from where the initial navigation map led, would be a suitable target.

Triumphant beetle. Beetles such as this are one of the most successful life forms on the planet, far outnumbering all humans. The one here has just snipped its way out from the inside of a wheat grain—the long snout contains blade-like cutters.

Why insect stings hurt. When the harpoonlike needle enters, a large chunk of attached muscle often breaks off from the insect, stays alive, and keeps venom pumping through the sting for up to a minute. Only female bees sting—the sting evolved from their egg-laying apparatus. Magnification about 60 times.

THE SECRET GARDEN

To land on.

And grasp firmly with your front legs.

And stick your head in to pollinate.

"Honey?!" she asks him, a little startled as he jerks back from her, but trying not to be offended.

Suddenly, jerkily, she moves back too.

Thereby flickering, and bringing the curious bee further down.

"Oooooph!" he gasps, glimpsing, again, this black-and-orange flying monster of a long-past childhood fear. He's scared of bees.

"Oh, darling," she whispers, pleased at what she takes to be his sensual exuberance and lack ofinhibition.

And so she abruptly leans forward, to hug her fella.

Thereby flickering again.

Which brings a certain eager flyer whapping even closer.

And makes her husband's eyes expand wider in yet greater horror.

"You're so . . . so . . .!" he pants, desperate to escape.

Arms pinioned though by his spouse.

"Yes!" she whispers, nestling her head against her hero-man's sturdy chest, barely feeling the little naked feet, hurry-stepping across the hair above her neck.

The bee doesn't have a great deal of time to explore, what with its limited honey fuel supplies running down. It keeps on waggling its front antennae to discover the concentrated source of nectar which it has been instructed is *somewhere* around here. And it moves its eye supports quickly, to try to make out the ultraviolet landing tracks, invisible to humans, which flowers generally have on their petals and which serve to guide a landing bee in. It will even sometimes try to rub in really close, to see if it can also pick up the slight electrostatic charge that an unvisited flower will build up by itself over an hour or two in dry air.

If you don't do anything, the creature will conclude that there's no useful nectar here, give a little running jump, and, with its wings throbbing at the distinctive deep 160 cycles per second—that's about the E below middle C—launch itself into the air. Then it will either survey again from a greater height, or just bank away and hurry straight back to the hive. Such a return is not at all uncommon: it's estimated that a third to a half of all bee flights end in failure to find any good source of nectar or pollen.

But if . . .

"Ahh!!" he says, louder, pushing free.
Slapping at the little creature trapped in her hair.

. . . you bother it, then, as soon as it has flown a few inches or feet up to safety, the bee will begin to pry open:

It's Stinging Chamber.

This isn't as bad as it sounds. Honeybees don't have giant hypodermic stings extending from their abdomen, which they use with careless abandon. Rather, their stinging chambers are a complex module, taking up a big part of their whole abdomen. There's the hypodermic, of course, connected to its small poison sac, but there're also muscular sheets, and various baffles, and a large section devoted to other glands, which release nothing more dangerous than a few billion molecules of some easily evaporable chemicals.

"Oh my God," he whispers, suddenly very still, as the small cloud of bees approaches.

The first of these chemicals, which is all the single lost bee has emitted so far, is only bringing in these other bees as a possible backup in case there's trouble. It might seem frightening as they all speed over, but . . .

"Why, it's just a few . . ." she says.
"Bees!" he screams.
"Yeah, well of course they're bees," she observes, looking at the little thing on her shoulder. "We're in a garden. There're flowers and stuff."
"LOTS OF BEES!" he screams.

. . . if you managed to look closely, you'd see that she's right, and they're really not doing anything dangerous. The emissary isn't going to do anything so foolish as sting a human just because it's been propelled up in the air little. Neither will any of the new arrivals. Bees die after they sting. And even the higher consciousness wouldn't want its emissaries to end their lives so fruitlessly. So the genetic instructions are clear and unambiguous. In these circumstances there's one thing that all bees will do:

They bluff.

They fly around in tight circles, and stick their front legs out in scary rigid thrusts, and dive-land to pull a hair and then take off again, and sometimes they even run around a little on their target before they take off again. And they lower their wing-beat frequency a little, to produce a scary rumble perhaps a half-octave below that usual E. But none of that involves stinging. It's just the usual posturing that animals engage in when they want to threaten another animal, and get it to back off, and yet somehow, get it to do so without having to engage in any potentially dangerous *fighting*. Dogs do it, with all their stereotyped threat-displays, and their just-as-stereotyped submission displays. Bees are no exception.

What most bee displays are designed for is *other* bees. Those, after all, are likely to be the most frequent assailants that honeybees have in the garden. If you're out searching for honey, the most tempting target is always going to be the fully loaded hive of *another* colony of honeybee somewhere in your terrain. That's the first reason bees have the whole paraphernalia of warning chemicals and threat-displays. But there's also another reason they have this, along with the stinging capability. This is because mammals sometimes attack hives too. Bears are noteworthy for it, and with their thick dark fur are extremely hard for the bees to get at.

The bees we live with, in response, almost all have a special visual sensitivity to dark colors, and to rough surfaces—what they would sense in the covering of other bees, and the furs of bears, and even the pelt-cladding of brave caveman raiders. This is why beekeepers are careful to wear white suits with smooth surfaces. It's also why bees will get most easily excited if you have on dark and rough clothing—avoiding something like brown corduroys really is wise. But still the creatures will be wary about going ahead and delivering that suicidal sting. They're intelligent enough, even as separate flying entities, not to lunge at just any dark rough surface. Too many such substances—think of falling tree bark—are not serious threats.

Bees have one more test hard-wired in their neural circuitry that they almost always go through before they let themselves go in for the fatal attack. It's an obvious one. You can get an impression of it in the surviving cave paintings from many thousands of years B.C., showing one especially powerful mammal that made brave and stealthy assaults on honeycombs to get crucially needed food for its own kind, ignoring the cloud of attacking bees we can see depicted in those paintings.

It's the species called Homo. Sometimes, humorously, further described as *sapiens*.

"If you'd only stay still," she says, chiding, not understanding his oddity, as a few more harmless bees fly near.

"Stay. Still," he gasps.

"And don't panic."

"Don't. Panic," he repeats. "Don't . . . panic."

Quieter now.

"See," she says, calmly. "Why, take this little one . . ."

The versatile wasp machine. The three domes (below) are on the head and work as a light meter to readjust the main eyes if the wasp should suddenly fly into thick shrubbery; the swiveling antenna base (right) lets it bend its antennae far forward, to tap cautiously if it has to walk along a dangerous branch.

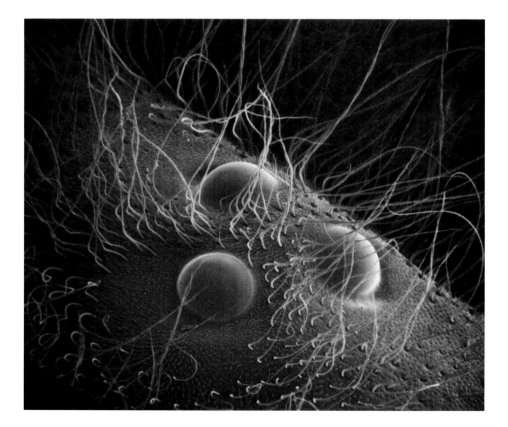

She looks sideways at another bluffing bee, which has landed and is crawling forward around her hair.

"AAAAARRRRRGGGGGH!!!!!" he screams.

"Mike," she says, almost exasperated at him—pleased, though, that he has managed to stay immobile.

And he looks at her. Eye to eye.

Trying, so hard, to breathe in a more controlled way.

"See?" she asks, kindly.

"Uh," he whispers, staring fixedly, as another bee walks out from his wife's hair.

"BEES! BEES! BEES!" he shrieks, flapping his arms wildly.

The moment of the sting itself is actually almost painless. The hypodermic on the bee's extremity is sharp, and goes in almost instantly once the creature detects—with its excellent flicker-detection unit—the sudden *extra* burst of activity down below. Even when the bee turns away from its target—tugs its wings at high speed to fly home

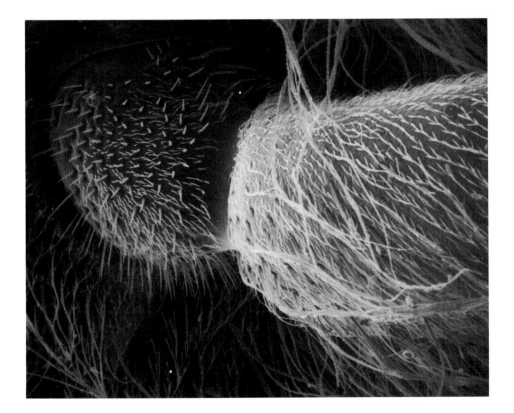

again, so pulling the entire stinging module from its own abdomen and leaving it projecting from the attacked human—there's still little pain. It's only after a few moments more, when the detached module's muscle-engines switch on, that you'll feel . . .

"THE PAIN!" he screams.

. . . something a little more unpleasant.

The bee's venom doesn't work by attacking you directly. That wouldn't be possible;

Loose plates on the lower leg (below) allow the wasp to bend at extreme angles while clambering, while the velvety suction pads visible inside claws (right) give it the grip to climb even smooth surfaces straight up.

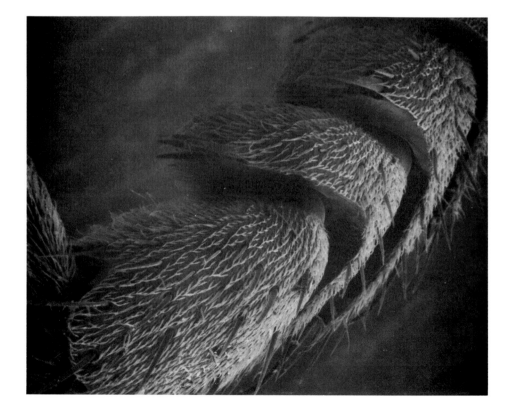

we're too big, and bees are too small. The total amount of poison the bee carries sloshing around inside itself is only a matter of milligrams, and much of that is water. The little bit of histamine which the bee has in there is only one percent of the volume. It might stop another bee, but is not enough to hurt a human, let alone an attacking bear.

What the venom—now being pumped through the quivering hypodermic—does instead is turn the attacked mammal's own immune system against *itself*. The main venom ingredient—50 percent of the total—is a protein that bubbles open the surface of many of our white blood cells. What's inside our own cells then comes out. Humans go around with great quantities of our own histamine, safely stored within those blood cells. It's supposed to be released in small, precisely aimed quantities, when the body needs it, as when there's a small cut or infection. With the bee's venom around, though,

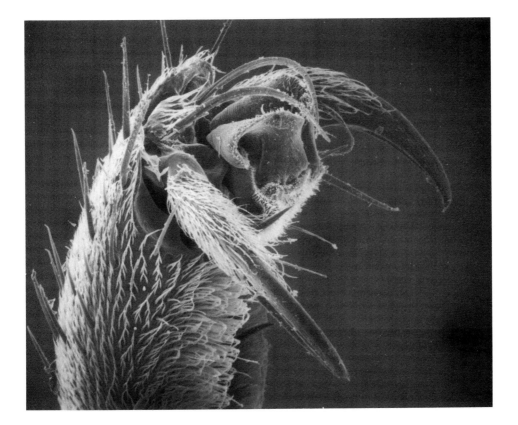

our cells just gush histamine. Much more bursts out than we'd ever release normally, which is why the stung area now begins to swell, and itch, and hurt.

Our immune system tries to defend itself. Connective tissues are quickly built up to keep any assaulting chemical from getting further in. Unfortunately, along with the first poison, the bees have injected a *second* substance—3 percent of the total volume of venom—from their poison sacs. It is a most unpleasant fluid. It works open the tissues which our body is trying to close to stop the spread of the first poison. The bee venom also contains another pain chemical, and a little bit of an allergy-triggering

The surface of the hindwing, close to the wingtip, is bristling with tiny hairs that contribute to the wasp's aerodynamic skills.

chemical . . . and so with the stinging module fast-pumping, and that second fluid holding burning tissue open so everything spreads, we are liable, even the bravest among us, to feel sore.

And then . . .

"Panic! Panic! Panic!" he shouts.

"Oh, do stop it," she says, finally exasperated.

"But it h-h-uuuuuuurrrrrtts," he whimpers, holding up one now slightly-throbbing stung finger for her to inspect.

At which another bee, one which may have managed to find the geranium, finally makes it over to the excited area.

"Bananas," she says. Calmly.

"Huuuuuurrrrrtts!" he repeats, insistently and louder, holding the finger higher, wanting a kiss to make it better.

The second bee approaches closer.

"Definitely bananas," she says, sniffing a little.

"H-H-HUUUUUURRRRRRTTS!!!" he pleads, screaming; all sense of humiliation gone; desperate now for that soothing kissy on the fingy.

Until he finally comprehends what she's saying. And then he sniffs it too.

As the second bee dives down to sting.

"BANANAS! BANANAS! BANANAS!" he screams.

. . . there really is the smell of bananas to consider.

Beekeepers know to back away from their hives and close them up, and maybe come back only a good few hours later, if they ever sniff it. The smell is produced by a gas called isoamyl acetate. This stuff is worse even than the second, tissue-opening fluid. All that did was tear open your cells to let awful poisons pour in. But in time that ends: the pumping hypodermic runs out of energy after thirty to sixty seconds, as the muscle engines run down. Isoamyl acetate, however, doesn't stop by itself. It's not even part of the liquid poison, but is produced elsewhere in the stinging module.

When the bee hypodermic gets jabbed into you, the module tears open a little, and the isoamyl acetate slips out. It's a light gas which evaporates easily, and so lifts away from the area of the sting. On hot days, when there's low humidity, the stuff evaporates even more rapidly than usual. If you ever smell it . . .

"Let's go," she says, teeth clenched, taking her husband's good hand, trying to cover the stung one with a light cloth.

He is speechless in his distress, but obedient.

. . . it is a wise idea to calmly cover the point of the sting, and walk slowly away. For the isoamyl acetate is designed to call in other bees. And when they smell it they don't just fly around and threaten and pretend. They follow it down to its source beside the throbbing hypodermic, and they *sting*. This is the second reason that waving your arms is so bad: not only does that goad the first bee into the initial sting, but it fans the isoamyl gas for the other bees to detect.

Luckily for us, the gas takes a few moments to float up—so we have that time to make our escape. It also disperses pretty quickly, if you're in the open, so it might well be gone by the time the other bees arrive. The human couple, following the sensible escape precaution, are likely to make it back to the safety of the house with nothing worse than a throbbing finger and a certain amount of retrospectively wounded dignity for one of them. As soon as the chemicals float away, the other bees will forget their threatening posturings, and re-form their intelligent skimming network in the air.

Quite unaware of the freeloaders who are about to use them to hitch a ride.

This is the final twist to the activities of the bees in the garden. It's carried out by the miniature juvenile form of a very common beetle (phoretic *Medoidae*). This tiny creature is usually known as a "triungulin." If you have strong vision you might just be able to see one of them as a black speck in your flowers. Dandelions and daisies are especially good to look in, for the triungulins stand out well against the yellow or white background, though tulips and others—including our geranium—often have them too. With a magnifying glass you're almost sure to spot some.

At night they stand still, inert in the cool air. But when the warmth of the day's sunshine reaches the geranium, they begin to perk up. Each one is an agile mountain-eer, with six limber legs. But a geranium is so high, to them, that if one ever plummeted off there would be an exhausting hours-long climb to make it back up. To guard against that, each triungulin secretes a little safety cord, something of a jump-master's static line, from the side of its body, and glues it down to the surface of its flower-home. Then if it is bumped hard—be it by a passing breeze, or any rough-fingered human doing the replanting—the triungulin won't fall all the way to the lawn. Instead it dangles safely suspended from its little ropes, the lawn still a vast ten-inch drop further below. Once the pendulum swings of its safety rope slow down, the micromountaineer clambers back up, winds in the excess rope, and reattaches itself.

Triungulins will do this any time they're bumped, but they reattach themselves especially strongly when they detect an emissary from the Distant Hive arriving. To a triungulin, that incoming bee—at first just a distant dot—quickly becomes a huge black shadow, grinding and buzzing its wings in an awful sky-filling roar, smashing up waves of air as it approaches.

Certain humans, by this point, would have long since passed out in fear. But the triungulin stays impassive. Because they're so tiny, they could probably survive, un-crushed, on the flower surface if they just crouched a little. But they don't do this. If you carry a single triungulin carefully on a glass microscope slide into a laboratory, and then pass a big looming shadow over it—your hand in the light of a bright lamp will do fine—it begins to stand straighter than usual. Bring your hand lower, increas-ing the shadow, and it stands even higher.

Now hold a heavy tuning fork, one buzzing and shaking at 160 vibrations per second, exactly at the speed of that E below middle C—just like the similarly dense roaring buzz of a hovering honeybee, in position for its final descent in. Then the triungulin jolts up on its leg tips into a grasping stance. And this is what happens in the garden. But what can a triungulin live on while waiting on the vast pastel-colored plain of an open flower? The nectar oozing up is too diluted for it to drink. What it needs is some concentrated, refined, *processed* nectar.

Exactly as might be found inside a distant beehive.

The closer the bee, the more there is of the roaring low E and that awful darkening shadow . . . and the more the triungulin arches up. Finally, when the bee is very close, when it's just about to touch down, then the triungulin flings the tattered end of its friction line upward, and itself leaps up, to transfer onto this great hovering behemoth from the sky.

How can the triungulin know to do this? How, for that matter, do any of these garden creatures know what to do? Hitching triungulins, and gas-communicating bees; earlier there were aphid-expelling roses, and rose-communicating aphids; microchip ants, and SOS-screaming leaves, and gigantic hidden soil cities, and all the rest. It's hard to see how evolution, working by blind chance, could really have created this extraordinary concoction.

So, once more, the evolutionary background:

Consider what happens when you leave your car at the garage over the weekend, and tell them to fiddle around as they think best. On Monday a man with a smile like the Luciferian Jack Nicholson in *The Shining* tells you there's been a little problem. Behind him, a cast extra from *Deliverance* looks up from stacking strangely familiar metal pieces—your transmission?—into a great bin. He plays with his backward-sitting baseball cap, and his name tag that proclaims him Gooper, and calls over something about the drill, oh, doggone it, come *on*, Elroy, couldn't ol' Gooper have a chance to fiddle around some more with the drill? You do not feel confident. Just so, you do not believe that flowers and leaves and ants could just end up doing these extraordinarily intricate things. You do not want these men, or any controlling deity in their image, loose on the DNA of your planet's life forms.

Evolution is simply a better garage. Many ancestral triungulins had mutations that didn't do much. But then a few, in times immensely long past, developed a mutation, an arrangement of electric nerve cells that responded just a little more than usual to the whap of a speeding bee's wing. That meant they got better access to the distant hives. That meant more next-generation triungulins. That meant the trait was carried on. And once that happened, once that step had become fixed genetically, "locked in," it didn't matter how long it was until other triungulins worked in the additional bee-shadow response, or the ability to clamber back up suspended ropes. Similarly with the first yellow-jacket wasps able to pick up that extra damaged leaf-alcohol.

Such success is not, except by unfortunate chance, going to be lost. Think of the garage owner not as Nicholson, but rather as a Ph.D. from MIT. He watches ol' Gooper kindly, and when ol' Gooper finally, by chance, connects the brake light to the right wire from the dashboard, he takes a Polaroid of it, puts that up on the wall, and instructs Gooper, and all the little, younger Goopers, to repeat it that way. Later one of the other Goopers drops a wrench on the engine casing. He's done that before, only creating dents, but this time it produces a mathematically unique hyperbolic curve in the combustion chamber that increases fuel efficiency by 2,160 percent. The Ph.D. puts up another Polaroid, and makes two phone calls. One to his broker, telling him to sell all his oil company shares. The other to a patent lawyer, saying: I think, sir, we have some attractive business for you.

Future historians stand with head-lowered reverence before that Polaroid, for it's famous now, in its atmosphere-controlled display case at the museum adjacent to the vast Gooper-Ph.D. motorworks, sited just beneath the heroic sculpture of the founder, with what historians conclude must have been the dominant late-twentieth-century head-covering style, that narrow cap with wide brim protruding in back, immortalized in bronze. So long as the right accidents are selected—and repeated—the technique is guaranteed to work.

That's why the triungulin can count on hitching safely aboard the quickly rising bee.

That's why it can count on being carried safely along the aerial navigation lines leading back to the Great Hive, even as the couple head in for lunch.

The geranium, and then the cucumber, the vegetable patch, and soon the whole lawn and garden quickly recede behind.

AFTERNOON

IN - HOUSE
VISITORS
AND
TALKING TREES

"OUCH!" he says, as she tweezer-pulls at the stinger, in front of the big plate-glass bedroom window, upstairs in their cool house.

"Shh," she says, concentrating in the light. "Almost there." She works the tweezer further in, to find the final bit of stinger.

"Yeeoww!!!" he cries, as she gets it, quickly jamming the sore finger into his trembling mouth.

It's pretty easy for our damaged hero to recover. Human saliva glands constantly produce tiny molecules of a most powerful hormone, which makes scratched skin surfaces grow together more quickly than they would otherwise. Normally that molecule (Epidermal Growth Factor) stays only in the mouth, where it can work on all the micro-scratches we get there from ordinary tooth biting and grinding. But anyone who sucks on a throbbing sore finger will get those molecules washing over the skin, and so help close the little scratches there. And if . . .

"Really, Mike," she says.
"Well, it h-hurts."
"I'm getting you an ice pack."
"Oh, great," he answers, with biting sarcasm.
But only once she's left the bedroom.

. . . more of the proper remedies are brought, the body will work on the rest of the problems. Coagulants are already being pumped into the blood, and a first scab is quickly forming, to keep what's inside from leaking out. Soon the blood vessels will begin regenerating, and the immune system will be sending out guardians against infectious invasion, and all the rest. Rather . . .

"Stupid garden," he repeats, enjoying the freedom of expression you get when you're alone.
"GODDAMN STUPID GARDEN: CAN'T EVEN TAKE CARE OF ITSELF!" he cries, waving his arms as he stands before the great window.
Really quite liking this revenge.

. . . it's the *garden* that should need some help in its recovery. For the rose stem was bruised, and the soil surfaces crushed, and the aphids were damaged, and the rest. It doesn't look like that strange human lunging at the upstairs window is going to help, so the garden will have to do the work on its own.

But that's no problem: gardens are terrific self-healers.

It's true that one of its parts is probably past healing. That bee which left its stinging module in the human's finger, and is likely to be staggering on the ground below the bedroom window, does not have much life expectancy now. But this is really not as bad as it sounds. Bees only live for four or five days as active flyers in our garden anyway: by the end of that time, their wings and insulating hairs are so abraded, their muscle

Inside a rebuilding chickweed flower. The female parts are growing out of the middle; leaning in from outside are the male parts. Yellow pollen dots are visible being transferred; the larger orange drops near the bottom corners are nectar for attracting bees.

stores are so run down, that they're not going to be much good for the hive, and they're programmed to die. The hesitant new flyers you see on Monday are scraggly veterans by Friday, gone by Saturday. The oldest bees are easy to identify, because the abrasion-baldness makes them shinier than the others.

The short life-expectancy of the bees solves another problem. Because the humans made the elementary mistake of spraying all their pesticide poisons early in the day instead of late in the afternoon, the flight patterns which the other individual bees make—this wondrous collection system of the aerial consciousness—has been turned into a machine for transporting poison back to the hive. But the individuals that form that machine soon die, and others take their place. The Hive wouldn't have survived this long if it was unable to respond to some poisons in the environment. The new individual flyers which take their place get new sorts of enzymes DNA-programmed into their body. When *they* take over the transportation lines, a week or so from now, their bodies will be neatly detoxifying the poisons, even as they're winging through the garden air with the nectar loads. (Though if you really overload the garden with drifting poison spray, even this correction won't work.)

And as to the rest of the garden, it's Growth Time.

Under the ground surface, the soil cities that had been hacked apart by the giant Hoe are rebuilding themselves. Crashed tunnel walls are opening, and the microcreatures that had huddled tight during the carnage are reemerging; even new miniature clay bricks are being pressed into position. Next to them, clover plants which had been broken off above ground by the same slamming Hoe are beginning to regenerate from their roots. They're even spraying out microblasts of the hydrogen cyanide gas they've been storing up during the morning attacks, and using it now to protect their tiny portions of turf against any new invaders.

A little higher, the crushed leaves of all the surface plants are slowly rebuilding, too. First the plant's internal pumping tunnels that feed into those leaves are reopening where they've been squeezed by crudely thudding Human Finger Attacks. If the thudding has been too great, or been followed by dreaded Fingernail Gouging, then fresh pumping tunnels are quickly made. Through all these tunnels come a tremendous range of chemicals. The first quickly refill the crucial ethylene gas canisters up there. Other chemicals supply basic structural molecules to fit into places on the newly re-growing leaf walls. Still others produce the defensive chemicals needed to repel the constant insect attackers, such as the blistery droplet fluid, which the geranium needs to keep the stubby red mites from spreading.

This all takes energy, of course . . .

"Hah!" he yells, from behind the big plate-glass window.
Taking in big, nostril-expanding pants of air, to help in his id-expressive vengeance dance.

. . . so the plants breathe more quickly during these exertions. Micropores on all the leaf surfaces open a little wider than usual to soak in more of the life-giving carbon dioxide which those wasteful human creatures may have left behind, after spraying it from their huge nostril openings. A special hormone is created and pumped up from the shallowest roots to control that plant breathing. If too much moisture is being lost as the air gusts back and forth, the plant is even likely to squeeze up a thin layer of wax, like a microscopic plastic wrapping, to keep some of the needed moisture in.

To keep resources from being wasted, the plants also have ways of registering, from the dark-inside of their pumping tunnels, what is taking place on their various extremities. Inside the rose, for example, the unusual gush of fluid streaming from its topmost pumping channels stopped as soon as the flower up there was snipped off. The plant registered that change, and rearranged its pumping system to supply new buds that are going to open.

The plants are also hurrying antibacterial chemicals through these dark tunnels to the points of the greatest woundings, to destroy any microbes that might manage to get in. In the case of the cucumber, not only is there a previously prepared gush of immune system–like chemicals as we saw earlier, but there's even a sort of booster-shot system. The plant sends out tiny doses of a *further* defending chemical, which will wait in place and only get copied and produced in great quantity a few months from now, when the first rush of protective chemicals has run out, and a backup defense is in order.

"Oh?" he asks, startled at something, not quite sure, for one long millisecond, what it is—only then realizing that in his fury against the garden, he's stubbed his bad finger against the plate-glass window.
"**#####*%***!!!" he comments.
And brings the swollen finger to his mouth.

Higher up, in the garden trees, certain small animals share this comforting recovery reflex. Where a tree leaf has been scratched by some errant clippers, there's likely to be a type of caterpillar around that can create a complex huge molecule in *its* saliva. The caterpillar presses its lips to the damaged leaf part, and—just as the human transferred his repair molecule from his mouth to his finger—the caterpillar transfers this huge molecule to the cut part of the leaf. The molecule makes the leaf grow back more quickly than it would otherwise—which is good for the tree, and also for the caterpillar, which needs living leaf tissue for its food.

This big-molecule caterpillar especially likes birches—the fresh-growing circles of green on the leaves are the giveaway that caterpillar-repair molecules are at work. But birches are pretty easy to survive on. What's more impressive is the bleary-eyed small beetle, shaped somewhat like . . .

"Again, Mike?" she asks as she comes in, seeing him standing there with the finger in his mouth once more.

At which he can only turn and smile weakly as she proffers the ice bag.

And brings out a small white tablet of compressed powder.

. . . a tiny oblong white tablet, mounted on equally tiny balancing feet, which manages to survive on the willow in the next yard. Life isn't easy there. Willows evolved to live near ponds or streams, where there's an especially high density of insect life. They've worked out some special defenses, a great many million years ago, for they wouldn't have made it to our time otherwise.

What they do—even the most peaceable of old garden willows—is bring together certain soil chemicals with their roots, and transform them into a vile-tasting poison. That poison is collected by the main pumping tubes within the tree, and then sent into the leaf or branch quadrants that need it. Any insects that land on this tree and try to take a bite there will end up swallowing some of that stuff.

With a little bite they still have a chance of survival. But if they ever snip out a big mouthful for tasting, the poison reaches the muscles of their stomachs and legs, and keeps those from tightening properly. Either the bug falls off, its legs frantically sliding as it tries to keep a grip—or it remains on the tree, but with its stomach no longer working, and so soon starves. There's no *chance* of competition between insect and tree.

Except from this bleary-eyed beetle.

You can often find these contented creatures on the hanging leaves of willow trees, solemnly chewing away. They're like one of those simple-minded eaters who go on *enjoying* their stew at the diner, while everyone else is racing out, or has fallen stricken on the floor, and you can already hear the siren of the ambulance. What *is* the problem? they say out loud, slightly befuddled at all the ruckus, before sighing in incomprehension, and going back for another brimming spoonful.

"It's not dangerous?" he asks, worried.

"This?" she asks, looking at the innocuous tablet, surprised that the medical book she's just consulted said nothing about profound brain disorder during the recovery process from bee stings.

"Yeah, well, maybe it's from Out There," her husband says, a strange gleam in his eyes.

"Mike," she says. "It's only . . ."

The reason the beetles can do this—the reason they survive where hardly any other bugs do—is that they can swallow the willow poison, and break it inside themselves into two parts. One is a simple nutritive food, which they live on. And the other part of the willow poison, the part which even these iron-stomached leaf beetles couldn't manage to digest, is safely poured into little scubalike canisters that these beetles have strapped inside their backs. Then when any threatening creature—such as the giant killer ladybird—crash-lands nearby, the happily chewing beetles simply face away, and open up the nozzles from their scuba containers. Out sprays that extra-concentrated willow poison, chemical formula HOC_6H_4COOH, and commonly known by the name of its main derivative, which is . . .

". . . aspirin," she finishes.

. . . aspirin: a substance excellent for forcing open smooth muscles in killer ladybirds, willow-marauding insects, and Human Beings alike. Since the inflammation and pain of a stung finger is due, in part, to our smooth muscles squeezing too tight, this willow-derived chemical is just right. Hippocrates prescribed willow bark as a pain reliever about 400 B.C., and it was partly from such traditional knowledge that nineteenth-century industrial chemists proceeded to refine the stuff, so producing the white powdered aspirin drug we know.

On the tree, the bleary beetles get a great advantage from this ability to tap the tree's own defense system. They're so soft that even the tiniest bite from a predator would be fatal. But with this willow-poison nozzle system they can keep those attackers away. In the first hours of life their reservoirs are especially well loaded, and the juveniles can spray dilute aspirin blasts pretty regularly for up to three hours without running out. As adults, some waddle on patrol through the willow branches with up to 5 percent of their weight composed of pure aspirin or related compound.

One of the only creatures that can get those leaf beetles off the willow is yet another species of small wasp, which swoops down, yanks the waddler by the *under*side, where there's no aspirin-spray nozzles . . .

"YES!" he cries, facing the window, detecting that rising blur.

. . . and once it has it a little way up in the air, starts banging the beetle's top *reservoir*-side against any exposed branch or tough leaf, to get the dangerous aspirin gas to fall loose. Though of course . . .

"Mike," she calls, pointing determinedly to the pill, and thinking: this is going to be a long lunch.

. . . if the selected beetle is crouching close enough to its fellows, as it often is, they'll spray enough aspirin up to interfere with the wasp's attack, and make it drop the first one.

Which means . . .

"Oh," he says, turning back, calm again, and popping the beetle-spray pill.

. . . that the wasp will have to try somewhere else for its midday food.

As the couple go to the downstairs patio for the lunch which she's carefully laid out, parts of the garden are likely to join them. For if you haven't showered or washed your hands properly after a morning outside in the garden, then a tremendous stream of garden goodies will be descending from your fingers, and skin, and hair, right down onto the waiting fresh vegetable salad below.

It's true that most of the billions of molecules of pine resin, and the residual bee-warning molecules, and the shrub's SOS chemicals will just evaporate up, and hover in place around your head or gust to the ceiling or the walls. You need to actually . . .

"This does look good," he says, realizing he's going to have to be the New Man to end all New Men if he wants a chance of eating.

He bends down to get a good sniff.

. . . throw your head forward for any of them to get right down there in the food. The pine resins can then give a hairspraylike fine coating to the bowl, as they had been

Bread mold. The original spores often float in as diffuse clouds from outside. Landed on bare metal or tiling they remain inert, but landed on exposed bread they quickly begin to grow. They probe into the bread to get nutrients; the small spheres are growth chambers for the creation of new spores.

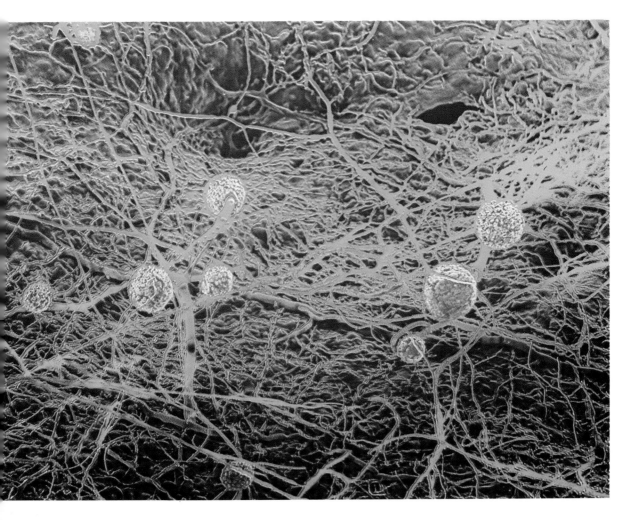

giving to your hands, and the SOS chemicals can float into the *other* vapors which the still-living lettuce and tomatoes are furiously puffing out, in their ongoing chemical warfare battle with each other there in the salad bowl. And . . .

"Yes sir," he murmurs.
Twisting his face in for a closer look, at his delicious honey-glazed salad.

. . . if you twist your head quickly enough, you're likely to fling down some of the early-morning lawn-climbing fungi, which have been desperately clinging to your hair strands since then. Those splat onto the salad too—falling through the gas battles of the living vegetables—and, as soon as they've oriented themselves, get to work trying to find some nice carved-in handholds to grip onto. An edge of one of the tomato slices will do well, though something deeper, like a tiny scratch on the inner surface of the wooden bowl, will be even better. Attached like that, the fungus can easily survive the meal. If this is a household where the salad bowl is only lightly rinsed after a meal, then it might well survive even that assault and get to sink its leathery arm-probes deeper into the wood, to draw nutrients from it and be ready for a Long Stay in this new home.

Now of course being flung through the air . . .

"Will you look at that?" he says, sighting something.
Poking his naked finger through the carrot slices to find it.

. . . isn't as guaranteed a way of arriving in a new habitat as direct transport would be. But it's good to remember that many of the objects you transfer onto your food when you touch it are quite innocuous. There will always be simple sweat molecules that have been produced in our body, as well as the oils, vitamins, mixed antibodies, and other objects that happen to be on the skin. Most of the bacteria you've picked up from the garden leaves or soil are just as innocuous, and simply sit, practically inert, on the surface of the salad vegetables once you transfer them across. Even the awful tetanus bacteria—which you likely have collected if you touched any hoe or pitchfork blade which had been in the deep soil—are almost certain to have died from exposure to the oxygen of the higher Upper World atmosphere in which we live. They're already inert, and perhaps have even begun decaying into the jellylike microblobs that almost all the soil bacteria on you will turn to in a few hours or so.

A few of the transferees to your food are even guaranteed to be beneficial. The many billions of fertilizer molecules you'll leave after handling a fertilizer sack (that's if you had gloves on; the figure without gloves is much higher) provides good, nitrogen-loaded food for the useful bacteria that live on the surface of your lunch food. The dismembered hollow spider leg tubes picked up from that stint of lying on the lawn are not quite so rich in nitrogen, but still provide good fertilizer—as do the tiny severed aphid antennae or leg parts that also are likely to have been in skin folds or under fingernails.

Not everything you may have picked up in the garden will benefit your lunch salad. That helpful stroking of the struggling geranium leaf didn't just suction up substantial numbers of the near-microscopic red spider mites, which have of course been frantically spraying out plant-control hormones ever since. It also, unfortunately, broke open a certain number of the plant's defensive microstumps, and suctioned up the blistering maze-acid which normally resided there. This acid is not especially good for the bacteria and other microbial life forms which live on your salad, and with a microscope you could see desolation zones quickly forming around any touch that brings the acid.

Neither member of the human couple is paying proper attention to what's flying up past them, also on the hunt for food, just outside their dining-room window. For from the time the sun came up, the air directly above the ground has been changing. Before dawn it was entirely stable, the ideal sticky sea in which the ants' curving walkways could anchor themselves. With the day's growing heat the air began to swell; it wobbled slowly, becoming a bulky invisible water balloon in shape, bending and stretching even though still bound to the ground. But finally, by this hour, just after noon, enough sun-spurted energy has been pumped in for the whole wobbling air mass to break loose. A great vertical elevator of heated air is lifting straight up.

Anything that happens to be at the bottom is caught in the air mass. Some of this is a matter of chance: the errant soil city creatures and the pudgy water microbears and the geranium's stubby mite-cars get no particular advantage in flying past the window this way, as many of them wind up doing. But it's not always chance, and a few creatures have evolved to take advantage of this powerful midday Great Uplifting.

The big-jawed bark beetle emerges from its deep tunnels in the pine tree, just yards away from the oblivious couple, and perches on the very edge, like a recently sleeping Superman on the edge of his high-cave diving board. The beetle senses the rising air bubble reaching its height on the trunk now; it gulps the air in deep, rhythmic swallows, then flings itself out. It falls at first, plummeting in the unfamiliar fight against

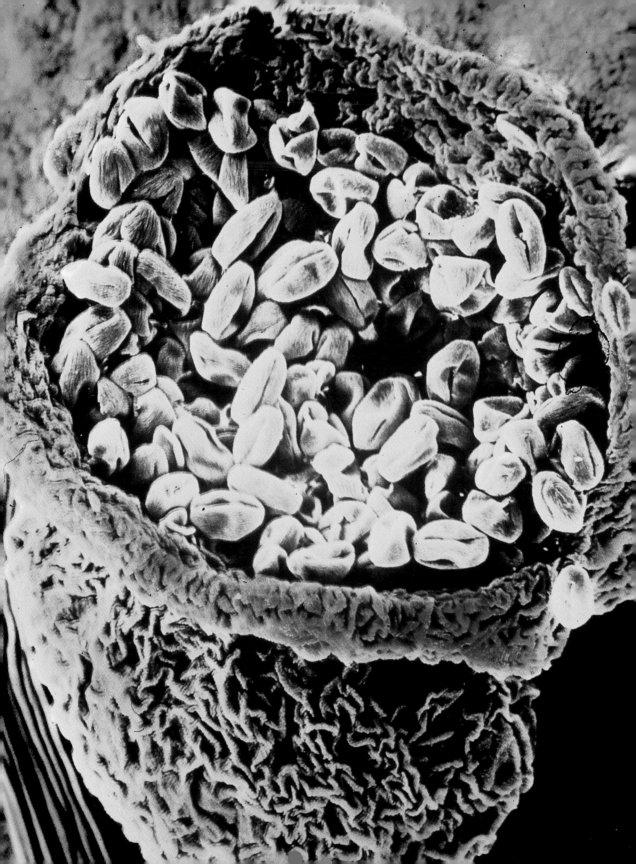

gravity outside the comforting dark tunnels, but quickly it gets up wing power and, aided by the automatic air elevator, manages to rise, sawdust-covered feet still dangling loose and purposeless behind, and just a path to the brilliant blue vault of the high sky in sight. At the right height the gulped air expands enough to make the beetle stop and do a slow-motion jackknife loop up there. Now that the bubble-altimeter inside it has brought the creature to the right surveying height, the beetle can rotate, and start speeding down, hurrying to another selected tree, to find its potential mate.

Joining the newly emerged bark beetles is that silent aphid community, which passed information among themselves to ensure that they would take off in time to catch this free midday shuttle. If you look you'll be especially likely to see a cluster of them here or there in the garden air now: tiny wings slowed down as the automatic air elevator blows strong and holds them high, with merely their ceaseless infrared surface scanning, searching for a good landing spot below, to show that they're still alert.

There are also the tiny optical spiders again, which greeted the human on the lawn surface before. If conditions are at all crowded, and there isn't enough food for their optical microhunting, then the youngest of them roll out great trailing silk lines now. On a cool day the lines just collect in clumps on the ground. But if it's hot out, and there's a good microstorm blowing, the silk lines stretch, and tauten, and billow upward in the invisible gale. Soon the microscopic baby spiders lift off in formation through the air beside your house: a miniaturized parachute regiment in reverse, also soaring up to the beckoning blue vault of the sky, to land, hours or days from now, where they can get a chance at life over The Great Fence, in the distant and perhaps more promising domain of other lawns.

It's a whole solar-powered flotilla, majestic in the soaring hot air currents. There are radiation-proof fungal spores, and vast numbers of living soil bacteria, and the roaring bark beetle behemoths. And there are breast-stroke straining fairy flies—creatures so small that ordinary garden air is as thick as water to them, and they maneuver in the up-blast not with wings, but with sticking-out feathery oars.

And then there's . . .

"Hnnngh," he asserts there at the table. "Uhhhn," he observes. "Haaa," he comments.
" 'Haaa'?" she queries.
"Haaaaaaaaa . . . Chooooo!" he concludes.
Pretty much on the colorful salad she's worked so hard to prepare.

An open pollen basket, showing the individual grains. Each pollen grain has buoyancy chambers inside, for air release when the basket opens. This basket is from Wild Cherry. Magnification 400 times.

. . .the pollen. This is a curious and remarkably toughened little object. Much of it flies up, but—from all the turbulence and sheer abundance—a certain amount careens sideways and gets on the patio and in the house. It doesn't make you sneeze simply because of the lunar-jagged casing it's covered with, nor because of the bulky dirigible-like air bags that are inside it, nor even because of the pesky habit it has of trying to mate with whatever object—be it Nostril or Flower—it happens to crumple-land on after its bobbing flight.

Rather, our at times unfortunately uncontrollable sneeze response comes about because of the sticky protein spread on the surface of pollen grains. It's there to ensure a good, mooring cable–tight landing. But such sticky proteins, at least for a large number of humans, also start a full-out histamine response—like a mild bee sting—leading to nose-twitching, eye-running, and ultimately the odd high-velocity burst of sneezing to try to get it loose.

If there were only a few hundred of those battery-powered dirigibles trying to mate in your nose at any one time, the problem wouldn't be too bad. Our natural body defenses would be able to swamp the few histamine responses those triggered. But on a midsummer day there can be tens of millions of pollen grains disengaging from your shrubbery and pumped everywhere by the sun-heat, including your way.

The small amount that's blasted out with the sneeze to land in the salad bowl isn't much of a problem, for there will almost certainly be so many pesticide-rich droplets traveling with it that there won't be many living objects left at the point of impact for pollen grains to attach to and bother. But if you're one of those people who finds it uncomfortable when these sticky surface proteins from the inside of a distant plant prepare for a bout of reflex mating behavior inside your nose, then the best solution is probably going to be . . .

"Isn't it incredible how, uh, colorful this cabbage piece is?" he now asks, smiling brightly.

He holds up that bright, red, honey-glazed cabbage slice he'd been hunting for.

Hoping, oh very much, that he can distract her from her anger over his sneeze into the salad, but realizing that he's not succeeding so well, when she doesn't say anything back; not even to argue.

She just pushes the salad bowl to his side of the table.

Takes the single serving of honey-glazed cabbage.

And expels it to the lawn.

. . . to go back inside, securely close all the windows, and wait till the floating pollen canisters already there slowly sink to the cool floor.

The more superficial members of the garden community do their best to get over in time to take advantage of this sudden cornucopia. Miniature rampways open, again, from the soil cities directly below the section of lawn that is receiving this Great Offering. Lawn roots nearby, and any flowers too, quite quickly detect the distinctive vapors from the honey glaze, and begin the motor-grinding process of directing their vacuuming root probes over in that direction. But the ants—no question about it—are going to get there first.

The first random forager who happens to be in the area only brings back a little bit of the stuff. But remember how these creatures were able to lay down those shimmering chemical walkways to mark out the path to a new find. In just minutes the garden ants will have created the mathematically optimal network for collecting the new food. It's something which the bees, stuck up there in vaguely marked aerial flight lines, can't do at the same speed or accuracy—and one of the big reasons that bees, though abundant, have to make do with being the *second*-most abundant multipart consciousness, achieving nothing near the overwhelming mass dominance which ants hold on our planet.

The reason the ants are so keen to bring back any sugary or sugar-coated foods is that they depend on liquid sugar to keep their miniature civilization going. In this they're like almost all the other small animals in the garden. Liquefied sugar fuel propels the flying wasps, and the staggering abdomen-gaping stinging bee, and even the tiny armored soil mites. It's a tremendous fuel: quick to go into liquid suspension so that it won't clog any supply tubes; easy to break apart; dense for its weight. It's because we share this chemistry of the insects that we share their appetite.

Back at the flat paving stone outside the back door, the first returned forager ant quickly steps down through the hidden bore-hole, leaving the Outer World as just a rapidly narrowing circle of light behind it. To keep from getting lost it switches on memory maps of this hidden Ant City. This is the equivalent of turning on fluorescent lights down there to guide it for its mission. It might stop to help dig out a caved-in worker—guided by the distress whistle which that worker produces from a transmitter on its abdomen—but otherwise it will push straight through, ignoring the crowds of *other* ant individuals, all hurrying on their own missions through the interconnecting passageways, till it finds the tunnel leading up into the one room it wants:

The incubation room.

This is likely to be right up at the top of the nest, where that paving stone forms a conveniently sun-warmed ceiling. There are rows and rows of waiting, snug-tucked babies there, and these are what the forager has been hurrying to feed. But of course . . .

"You know, I could handle some good burgers now," he calls out, hopefully.

In awful hunger-agony as he hears her, through the swinging door to the kitchen, grilling them on the charcoal hibachi, with its specially installed extractor fan.

The preposterously calorie-empty salad bowl sitting mockingly on the table before him.

. . . it won't feed them as much as they want. Ant babies that were fed on demand would become behemoths, giants two or three inches long, and so could be tempted to escape, smashing their way out from the enclosing tunnels. Nor will it give them the ingredients . . .

"Nope," she calls back, simply.

Then takes a juicy, spattering, nostril-compelling burger, and places it, deliberately, on a big white plate.

For herself.

. . . that could make them capable of becoming fertile males. Only a few sexually capable males are ever produced in the Ant Civilization, and usually only for those brief intervals when the Queen has need for them. After they've performed their single utilitarian function, the males, due to an unkind genetic programming, find that their massively swollen wings lose hydraulic pressure, and soon fall off; their muscles quiver, their whole bodies weaken, and then, often preceded by . . .

"Aaaa . . . ," he sighs.

Dropping his head onto the table.

One closed eyelid trembling in desolation.

. . . a pathetic rolling over onto their backs, they die.

Instead, almost all the babies are created and fed so that they turn out as small sterile females, closely related to each other. Energetic entomologists have reached into nests to collect workers of a fixed age, and have analyzed their DNA. If exo-entomologists from Sirius reached into a human-populated shopping mall, and lifted some of us up for analysis, they'd find that human brothers or sisters have about 50 percent of their genes arranged in common. Among the ant sisters it's different: they have 75 percent of their DNA arranged in common. This means that more of their DNA survives if they stay at home and take care of their sisters than if they flee the nest and try to start up on their own, producing daughters. It's even worthwhile for them to die for the sake of their sister ants, because more of their genes are carried on through the sister than would be if they let *her* perish and produced a daughter of their own instead. This is why almost all ants you see are female. It's a guaranteed basis for altruism.

Even with the reduced food-supply levels the protofemales will need, there's still a lot of babies in a small ant colony. The foot-wide city excavated in your garden can have several hundred or more. To feed them all the forager ant needs more energy than even a mysteriously landed great red cabbage slice, however attractively glazed, can provide. That's why it quickly leaves the nest, and ascends into the Upper World again, to find some additional sources of sugar.

As picnic-goers can testify, the ants' next stop is likely to be one of the garden's trees. But this can present problems, even with the convenient above-ground tunnel network stretching there. Because of the noon surface wind, lots of bacteria capsules are whirlpooling up and impacting on the walking ant-machines. Without some protection these capsules would activate, and live bacterial creatures would soon be out hunting on the creature's body. Luckily the ant got coated with a few antibiotics before leaving the nest, and also carries a mobile storage container of more antibiotics. The stuff is stored in a liquid form, in a container near the ant's waist. When it's splashed on, the antibiotic liquid dissolves most of the storm-blown bacteria which get stuck on it, and incidentally helps give the ants their distinctive glisten. It's a bit like what happens in humans. A certain amount of antibiotic liquid coats our eyes all the time; when we cry, and there's an increased chance of gasping in some bacteria, extra amounts are ejected from the production gland under our eyelid into our tears.

The ant also has to make do with the problem that its chemical odor walkways are likely to be floating up away from it. Lawn-base air is sticky, but it's not that sticky. With a full microgale blowing, big sections of the translucent tunnel snap off and lift away into the air. The ant has to find it. At first it just forlornly sways its antennae to both sides—as we might anxiously glance around, trying not to be too obvious about it, when the unpaved road we've been driving on suddenly runs into a dirt path. But then its genetic instructions lead it to a more efficient sweep pattern. It pokes its antennae carefully high, to keep as much contact as possible with the rising walkways as they lift; it waggles its antennae carefully low, to pick up the fragments which might have remained scratched in the soil. Eventually it picks up the conveyor-belt pathway again.

Finally at the great tree, the tiny ant begins to climb. At its small size, gravity is not a great problem, and the ant can usually go straight up without slowing. There will be landed fungus spores on the furrowed bark it climbs, some even with the live fungus creatures slowly extending out to see what this landed world is like. But the ant obliviously crunches these bare capsules underfoot—new leathery fungus arms splattering—in its rush to the target above. A few hesitant beetles still in cavelike holes in the bark might extend their heads outside at the commotion, but the fast-climbing ant stomps past them too. Only when it's higher, two yards or more above the ground, where the tree's branches begin, does it stop the fast vertical climbing and take one of these bridgeways going horizontally now. One twig end is investigated, but WHOOPS, there's something Big there already, so the ant turns back and races higher, trying yet another twig, hurrying along to a farther leaf. Finally it reaches its goal, its crucial fuel supply and the salvation of the now-distant ant city:

The Aphid-Crawling Leaf.

It shouldn't be a surprise. Trees easily support thousands of aphids—2.3 *million* have been counted on one large sycamore—and aphids are the ideal creatures to be supporting things. Since their escape from the rose and cucumber this morning, they've been cruising high in the garden's air currents, constantly scanning for a place that's neither too hot nor too cold to land. The high canopy of a garden tree is an ideal place for them. One tiny aphid trying to pull sap up the length of the tree wouldn't be strong enough to start the tree's inner circulation tubes. But aphid creatures, with their uncanny synchronization ability, rarely have to hunt alone.

Instead, as the flotilla of tiny pumping machines landed, they formed into position, clustered in a big family herd as always—feeding tubes down and rear ends up—and then they All Switched On Together. That sudden jolt jump-started the leaf. Once installed, the wrinkly aphids also synthesized a large molecule, almost the same as the one the caterpillar made on the birch, to keep the tree sending its sugar-rich inner sap up. Again there's an advantage in numbers: several hundred aphids will be pushing this molecule down through their pumping tubes at the same time, into the tree's circulation. The tree seems to have no alternative but to gush back the nutrients which the aphids are now controlling.

It's a remarkable skill, and of course it demands a certain manipulation of the other creatures that happen to be around. The ladybirds, as usual, are a problem, and up here there's also a danger from the constantly circling wasps. This is where the ants, quite casually searching for the sugar fuel for their ant city yards below, come in. When a giant air-flapping wasp or ladybird tries a landing on this exposed upper canopy, ready to collect some nice aphid meat, one forager ant will walk over, widen the nozzle of an air-dispersing poison gas it carries on board, and let loose. If the wasp or ladybird is huge, and the ant is small, this is suicidal. But the other ants hurry over. They might be too late to salvage the first one, but together, all spraying the poison gas, they're likely to be able to drive the landed flyer away from the aphids.

In return for this assistance, the aphids let them . . .

"Hey!" the human whispers to himself, remembering something.
Taking a sip from his sugary Coke, as he contemplates the great discovery.

. . .sip from the excess sugars which they spill out as they draw them up from the tree. Quite a bit comes out—in Australia, skilled aborigines have been able to collect up to three pounds a day from a range of trees. This procedure has not always been historically obscure. There's an aphidlike creature on the tamarisk tree in the Middle East which also pulls up excess sugary tree fluid that has been collected. The Arabic word for this sugar-liquid is *man*. It could well be what at one time, and under the more elaborated name of *manna*—see Exodus 16:31—was collected a little more religiously.

If aphids really are in control of all this—leading the ants around to protect them; switching on tree leaves at will—they're going about it pretty oddly. Why are they only on the leaves at the exposed tip of the swaying tree canopy? The safer leaves farther in they don't touch. Any true Master of the Universe, let alone Lord of this Garden, would say Halt, Avast thee, Oh sinful inadequate tree leaves, and let me get back to those inner portions of the tree, where all is leafy and green, and, hey, a whole lot safer too, with no great wasp helicopters to attempt grievous damage unto me. But this the aphids cannot do.

Somehow it's as if the tree were *intentionally* luring the aphids over to the outer branches now, where—despite the partial ant protection—they could be preyed upon by these landing wasps and ladybirds. And when you realize that trees pump in aphid-attracting water to the weakest, most easily replaced leaves, precisely on those outer portions where the lured aphids can be harassed by the wasps; and also that those leaves are supplied, by the tree, with just the loose nitrogen the aphids toil for . . .

Glucose—the universal energy source. Abundant in honey, sap, tree leaves, aphid bodies, and human blood. Chunky crystals are usually far too small to notice. Magnification 250 times.

THE SECRET GARDEN

"Time for your television show!!!" he suddenly calls out, which is what he remembered, his watch beeping 1 P.M.

"Oh God the remote control!!!" she says, sprinting from the kitchen, needing this device to catch her favorite soap opera.

"Upstairs!!!!" he shouts.

"Got it!" she calls, racing up there.

"Good," she gasps, finally settled in front of the TV now.

. . . then it's clear that it's not the feeble aphids who are really in control.

It's . . .

"Great!" he says, now having made his way to the vacated kitchen.

Triumphant at having manipulated her out of the way again.

Limitless hibachi-meat his for the taking.

. . . the *Tree.*

To understand this, we have to discard any notion of trees as big static clumps of lumber and leaf. Rather, look at them as busy, constantly active vertical factories. What has only been recently recognized is that these trees have the chemical ability to register in just which quadrant of their canopy there are insects that could be dangers to them. Even more, trees then selectively target these regions, and spray or pump zapping chemical right to the point where they're needed. There's no "intelligence" in this of course—trees have no brain cells—but just a good evolutionary fitting, as we saw with ol' Gooper in Chapter Three. The tree gets enough energy for all this from its leaves, and because it has roots stretching everywhere—under the lawn where the microchip ants navigate, into the vegetable patch where the cucumber is losing its poison-gas battle; even to the transplanted geranium, crunching open even more soil cities now as it drinks and feeds. It's a giant conveyor belt, and all the water and raw chemicals needed for the cunning defenses come pouring in.

All garden trees have the power to do this manipulating, though they do it in different ways. While the oak directly shifts around the aphids on it, the neighbor's willow manipulates its aspirin beetles more subtly. It's not just a matter of putting out too little of the precursor chemical which those beetles need to make their aspiring gas. That would only reduce the beetles' population. Rather, a willow tree chemically senses where there are too many attacks by *other* bugs, and pumps stores of this precursor chemical into its bark there. The bleary-eyed beetles have to follow it. They become a mobile patrol of chemically well-armed sentries all under the willow's control. Once they've concentrated in the threatened portion of the tree, they quite naturally spray out the aspirin gas to defend themselves . . . with, as a most convenient side effect, the willow getting defended along with them.

In pine trees, something similar happens with the thick resin which the tree pumps to supply its growing needles. In the early part of the year, when there aren't too many bothering insects around, the tree pumps that resin to the central needles, well protected in the largest branches. But as the season goes on, a small hovering creature called the sawfly hatches its larvae on the pine. They love that nutrient-rich resin, and if all the resin just stayed there, in those safely protected central branches, the tree would have little chance of getting the sawflies off. What can it do? The SOS chemical which the pine released to call in predators of the larger beetles burrowing in its bark wouldn't work here, because it would become too diffuse, with all the tiny sawfly larvae out hunting. Instead, quite cunningly, the tree begins to pump some of its resin away from the middle and out to the *outermost* needles. The sawflies have to follow their food. Passing wasps and birds grab a few sawflies. The pine pumps more of the resin out. More sawflies follow. More birds get plump . . . and more pines get saved.

"B-but . . ." she dimly begins, standing up from her seat in front of the TV.
Realizing he's now alone, in control, with all the meat-food.
"It's starting!" he calls out.
So sending her back to facing the set.
Obedient once again.

Often trees don't do this shuttling to bring an insect to any specific exposed killing ground, but simply just to keep the insect *moving*. Caterpillars especially suffer from this. An oak will load one big section of its canopy with the water and nitrogen those caterpillars need, and then, even before they've finished their fastidious feeding, will pump those first leaves a little bit drier, and send the nutrients gushing over to other leaves, new ones, yards away in the canopy. The oak, in this response, hasn't had to work out where exactly the caterpillar is, but just does it on the good chance that *some* bug or other will be moved this way.

The poor naked caterpillars have no choice but to release their stabilizing tripods, energize their wriggling legs, and try to crawl across the branches and twigs to reach that new feeding station. This forced migration is excellent for the oak. A crawling caterpillar is an energy-depleting caterpillar. It's also an overexposed caterpillar: the garden's birds now get the chance to use their optic-movement detectors and pick off these slow wriggly herds.

Even if the caterpillar herds managed to avoid the prowling birds, by excellent camouflage or perhaps by traveling only at night, the oak still gets an advantage by forcing them to crawl from spot to spot this way. Think of everything the nimble-footed ant stepped over on its rush up. Caterpillars don't have such high-stepping legs. They're neither nimble nor skinny, but simply fat and short, and get exhausted when they try to go any faster than their usual soft waddle-crawl. They have to push through tracts of bacteria and fungus particles which have crash-landed from the garden air onto the tree's bark. Some of those get zapped by poisons which the tree oozes up to defend itself, but often enough, for species of fungi which are good against passing caterpillars, an oak will ooze up not poisons but excellent fungus-nutrients to *feed* these helpful landers. The caterpillar, waddle-hurrying forward, head low above its stubby legs, has to open its mouth at some point, and so inadvertently gulps them in.

Of course sometimes . . .

"It's not on!" she calls out when desperate channel flicking shows nothing right.
Dangerously restless.

. . . a garden tree has the misfortune of being bothered by one of those many bugs which cannot be defeated by aspirin, or led around by nutrient shifts, or picked off by birds. In that case it has to fall back on a quite different way of shaking them off. It doesn't hide from these attackers . . .

"Mike," she calls ominously.
Looking to the still-swinging kitchen door.

. . . by simply cowering in space. That would be neither suitable to a garden master tree, nor . . .

"I know you're in there, Mike," she says, an unpleasant note of determination rising in her voice.

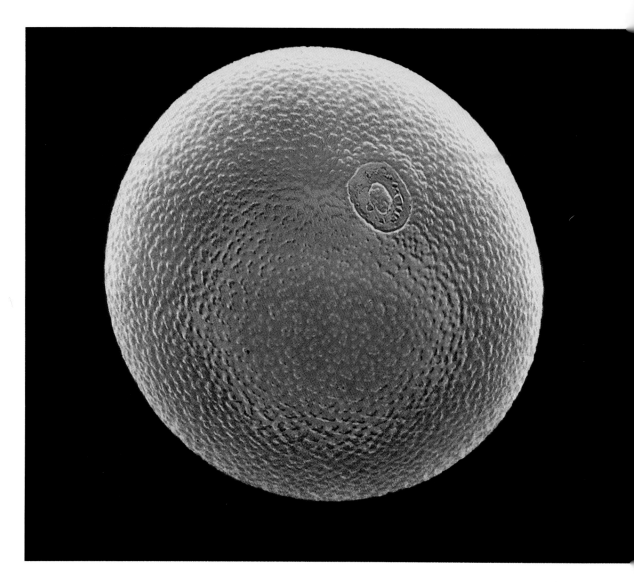

. . . effective: How could a giant solid-wood structure "hide" from a persistent attacker? Rather, the intelligent tree . . .

"M-olp," one anxious man, hiding in the kitchen, manages to stammer.

. . . works out a way of hiding in *time*. For normally the bugs will have evolved to have hatched at just the right week to catch the tree with its softest buds just beginning to

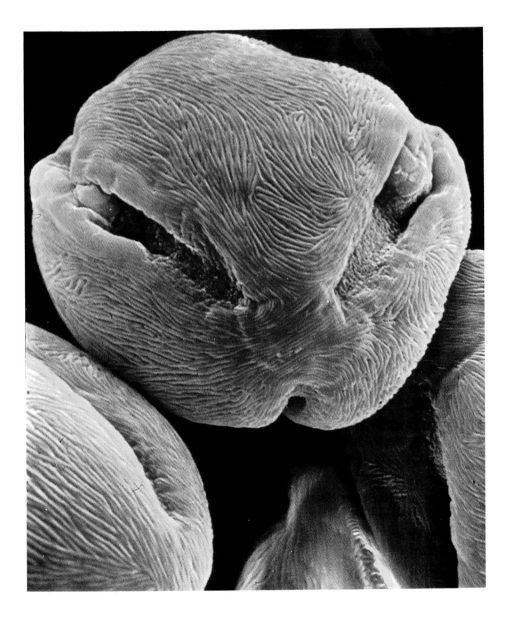

Pollen grains, from a pear tree (below) and wild grass respectively. Sticky proteins on their surfaces help in attachment when they land. The furrowed "eyes" of the pear grain are exits from which the male reproductive cells will later emerge; in the grass pollen they'll emerge from the already visible dot in the "dimple" near the top.

open. But an intelligent tree (and this will include the garden's oak, or a maple, or indeed almost any other ordinary specimen) . . .

"The football game!" he suddenly calls back.

. . . takes this into account. It stops budding at its usual time in the spring. Instead, it adjusts its internal machinery, so that the bugs can't be sure of "catching" the next springtime rendezvous. Sometimes it'll bud a little ahead of time, so that it'll be grown enough to be able to defend itself when the hatched bugs come up. Or sometimes it'll bud a little later than usual, so that the hatched bugs find that they've arrived on a sterile, food-empty world, and starve before the soft new oak leaves come out.

"The . . . football game?" she asks.
"It delays all the shows," he says, poking his head around the door, thinking fast. "You have to wait. I can't tell you exactly when."
"Oh," she says, understanding that.
She goes back to sit . . .

This "time-hiding" defense is limited only by the fact that the trees can't start *too* much earlier in the year—then they could freeze in a late frost. But even a variation of two or three weeks . . .

. . . for a while.
While he gets to finish making his food.

. . . is often enough.

It works after a fashion, but you'd think something better should be possible. Shifting an entire budding system through weeks of time just to avoid a few tiny bugs isn't really efficient. Similarly for sloshing hundreds of gallons of water and nutrients from one side to another. Wouldn't it be better by far if your tree had greater finesse? If it could sense *exactly* where those troublesome bugs were . . .

"Hmm," she murmurs, bored at the game.
She reaches for the remote control.
To Zap it.

. . . so it could Zap them?

This, as it turns out, is something the oak outside the dining room, and most other trees, do almost every day in the summer. When a really determined twig-walking insect bites into one of their leaves, small fragments of shattered leaf edge fall into the center of the damaged leaf. There the leaf pieces start the DNA motor-controllers inside —and from that point the tree has the needed signal showing that there's an attack precisely there.

Having the pieces activate the leaf they were torn from is a clever location-device, because it's almost impossible for any biting insect to avoid triggering it. No matter how quiet, how careful the bites that the bug takes, it's still a bite. Those motor-starting leaf bits are going to be let loose. The Klaxons blast and the bells ring and the search-lights crack on and the brass band starts up. It's the same thing that happened to the cautious webworm caterpillar on the SOS bush earlier. There's no way to break open a leaf without releasing such fragments. For either . . .

"Chewy," he remarks, as he finally enters the living room, biting into his nice double-burger with extra garlic.

He settles down alongside his spouse.

. . . you chew, or you don't chew. These are the basics of mealtime.

The fragments don't merely slip into the nearest leaves, but get swept up by the tree's huge internal circulation system. In a matter of hours they reach nearby leaves, and switch on *their* DNA motors, too. And what they produce is awful: it's the same tannin glue which the rosebush used earlier against the aphids. The tannin is pumped back to the activated leaf, and any caterpillar caught on the leaf is going to find its chewing slowed, and its jaws seizing up, and pretty soon its stomach turning into something closely resembling well-tanned leather.

Nor is it any good for the caterpillars to try to hurry to an adjacent twig. When an oak's sensors locate one of these caterpillars, it doesn't just send the glue poison back to the first leaf that's been attacked. The poison gets pumped back to *all* the leaves in the vicinity. A caterpillar sampling any of them suffers the same fate. Unless it man-ages to transform itself into an air-flapping moth or butterfly in time to escape from the tree, it will end its life, quite isolated, there in the localized green desert of suddenly glue-tinged oak leaves.

A few garden insects try to make do by getting off the leaves entirely and hiding on

THE SECRET GARDEN

their tree's bark. Yet the insects don't automatically survive, even if they do manage to get off those strangely "alive" leaves. In the case of balsam, scientists have worked out a macabre story. Those trees build up, inside themselves, complex molecules the design of which they've "stolen" from insects over the millennia. The tree oozes this molecule out through its bark, and the insects, trustingly hiding there, end up breathing or eating it. But once they do that, something happens. The insect gradually begins to "stretch," almost as if it were trying to grow out of its own body. Its limbs alter, and the whole creature swells horribly. What's happened is that the molecule the suppos-edly "dumb" tree produced is a near duplicate of the insect's own sex hormones. But the tree gives the insect the wrong doses, at the wrong time, and turns the insect into a monstrosity.

This molecule was so unsuspected—what's an insect hormone doing *inside* a tree? —that it was only discovered when Karel Slama of Czechoslovakia tried to raise his usual supply of research insects while visiting at a laboratory at Harvard. Although Slama had managed it with no problem back home, something was wrong in America. He and others traced the problem to the paper towels in the growing jars. The ones at home had no balsam pulp in them, but the ones in America did . . . and it was in that balsam that the first of these extraordinary age-distorting hormones was found. Other hormones and even stronger substances have been found since, all produced in these wooden chemical machines.

As a last resort, some garden bugs try to stay on your trees not by poking around in the bark or any other potentially lethal place, but rather by sticking tight to the first leaf they land on, and not budging at all. The way they try to protect themselves on this last stronghold is by snipping carefully into all the thin exit tubes which stretch from the leaf into the main twigs and branches of the tree, so that the usual alarm system is shut off. No warning signals can be sent out, and no awful tannin glues can come gushing back in. If you look around, you can see the caterpillar form of the

An ordinary sodium compound under polarized light in the microscope. Sodium is central to controlling fluid balance in cells. Sodium compounds change their function drastically with a simple change in what the sodium is joined to—thus sodium carbonate (washing soda), sodium cyanide (poison), or sodium chloride . . . which is just common salt.

common monarch butterfly carefully cutting away in this fashion. Other creatures—look out for some of the most common beetles on cucumbers—do a similar trim. And if you look really closely, you'll see that the insects are careful not to snip *all* the connecting tubes, but leave one or two, so that the gush of nutrients which the plant normally pumps into its leaves can continue.

Sometimes the plant can't fight back, and the bug's strategy succeeds. But sometimes on a big tree, something most destabilizing happens to the leaf which the ingenious insect thief is blocking off this way. An oak might sense this attack, and begin to puff out little bursts of that gasolinelike warning vapor—that ethylene gas—we saw earlier. It leapfrogs *over* the blocked part and starts making the leaf loosen. For the insect this is awful. Its entire leaf-island world begins to wobble and shake underneath it. For the tree, of course, it's a sensible gesture: far better to get rid of one replaceable leaf than to keep it attached as an open spigot, with the insect on it steadily draining what comes up.

The quick . . .

"Phew!" he puffs, in excitement, as the soap opera heats up.

. . . puffs of ethylene get stronger, the leaf begins to wobble even more, almost loose, almost sending the insect plummeting to destruction—but then, just before the leaf falls, the tree's internal motors switch into reverse, and it sucks back the last bits of nutrients contained in the leaf. Only then does it give the final puff of ethylene gas to cut the leaf clear, and send it circling, in free-fall, to the ground below. Only then will the tiny tubes the tree has readied, sticking *up* from its topmost roots, prepare to go into action, and, as happens to tree insects almost every time a leaf gets flung off, . . .

"That's a jungle," he declares, in cathartic shock, as the credits finally roll.
While his wife, numbed by the on-screen awfulness, can't even speak.

. . . eventually slurp into themselves the fragments of the disoriented insect aeronaut, at this terminal end of its flight, just a few yards outside the peaceful house.

A final step. The ethylene warning gas is good for cutting off damaged leaves; it's also, as we saw earlier with the rose, good for triggering a plant to produce more of the sticky tannins that are needed for general insect-defense. Why then should it stop at the outer canopy of one individual tree? If it could float on, and alert the neighboring trees that there's an insect attack coming up, then *those* trees, too, would be readied in time. Naturally it would take a lot of the warning gas to get a second tree to switch on its tannin motors this way. Only if there were a good number of insects working on the first tree, and so a good chance that some of them would be soon heading over to the second tree, would this air-warning—this invisible air-*talking?*—come into operation.

Certain evidence has come out recently that, at least for some trees, this actually happens. The released vapor has been measured, and the molecular sites on tree leaf cells where it lands and leads to the switching on of the tannin-production motors has been identified. In young poplars, sugar maples, and willows, it's been possible to get some accurate timings on the invisible network of aerial tree communication. About thirty-six hours after a caterpillar attack on one young maple, there'll be enough of this ethylene warning gas sent out to start the tannin production on nearby maples, and get them ready for the advancing bugs.

The trees we see, standing so apparently quiet and isolated, are actually hooked up by these invisible gas connection lines, working and defending in unison.

To the forager ant, now finding its pathway back amid the elevated leaf islands, and to all the other garden creatures, all of this is invisible. The tree the ant is navigating along is just a great wood-creation machine, which somehow supplies aphid pumping stations. But the ant, of course, lives on too quick a time scale. To the slow, ponderous tree, the ant, the aphids, and all the other garden creatures are barely visible blurs; preposterously speeding life forms. Even the television-shocked humans, in the safety of their brick-and-stone construction far below, are just strange fleshy bipeds, squeaking air vibrations to each other as they hover for what seems only seconds in front of the cooling television, before engaging in some enjoyable gymnastics in the bedroom upstairs. Even when one finally scurry-emerges from their home, plopping down to assume a stable position, holding up a reconstituted sheet of wood fragments to read, this massive, ethylene-gusting, slowly communicating oak he's leaning against remains above all that.

Think of a massive boulder, untouched by the sea-spray, jutting up impregnably before the ocean.

If we could see the world at the tree's ponderously slowed rate, that's what the tree —this Garden Master—would look like, in control, now.

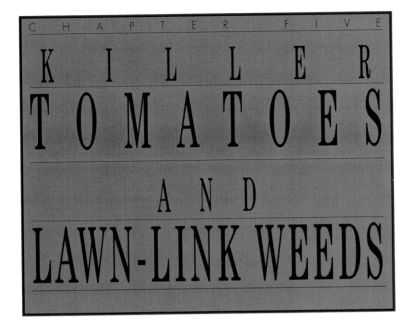

CHAPTER FIVE

KILLER TOMATOES
AND
LAWN-LINK WEEDS

Since it will be a little while before the soil cities manage to disassemble the leaf-flung caterpillar into a form that the tree's great roots can digest, the oak has to prepare its own meals for now. That means, first of all, suctioning in some . . .

"It's a bit dry-y," he tries sing-song calling to his wife, who's still inside the house, collecting something from the herb box at the open kitchen window, as he begins eating that new Lo-Cal plain cracker they're trying for his post-lunch treat.

. . . water from the wet microtunnels in the nearest soil cities. The tree needs that because all the food it has stored inside itself, all the millions and millions of tiny . . .

"Starchy, actually," he reflects, leaning against the tree, newspaper slightly to the side, as he holds that cracker bite in his mouth.

. . . starch-bubbles in its gnarled roots, are no good as they stand. The starch in plant roots, as in the bread and plain crackers we humans eat, might be one of the most abundant chemical forms on the earth's surface, but it's simply sugar that has had its water molecules chemically clipped away. That makes it good for storage—you save space without that extra molecule—but to switch back into food, starch needs to get those water molecules squeezed back on. Simply swirling starch and water around won't do the job. They'll stay separate. What you need is . . .

"Well, maybe not that starchy," he thinks, as it strangely begins to taste sweeter.
As his saliva molecules roll all over the cracker.

. . . yet another group of specialized molecules, such as exists in human saliva, which *can* squeeze that ordinary water onto any waiting starch, and so turn it into the sort of nice sugar hungry stomachs crave. It turns out that oak trees have almost the same specialized molecule, and use it when their food needs are running high at midafternoon. In fact, when an oak has been suffering strong insect attack, and needs to pull up more food to build its gas and chemical defenses, then . . .

"I'll try another one," he thinks, taking another bite, as those saliva molecules flow.

. . . just as our saliva glands do, its roots quickly produce even more of this digestion molecule.

The rest of the garden can't let this go on without some struggle. Otherwise the tree, with its great length of powerful hydraulic-roaring roots, would grab everything. So from the dangling root bottoms of the grass, and flowers, and shrubs, desperate gusts of . . .

"Now a sip of coffee," he thinks, but makes a face when he takes it.
"Gosh, that's bitter."

. . . bitter tannin or tannin-related gases get sprayed out as the gigantic tree roots push near.

To some extent this works—gardens aren't taken over by oaks—but it doesn't work entirely. Your oak acts a little like the squabbling cucumbers and tomatoes in the vegetable patch, and slides open molecular hatches in its roots to swallow much of the incoming gas. But remember that the tree's roots can be curving around for many thousands of miles in all the stacked volume within the hidden soil cities. There is a terrific amount of dangerous tannin-related gas being poured into that area from the other garden plants. An oak can't store all the digestive chemicals needed to neutralize the incoming poison the roots are swallowing. Instead, on a sunny afternoon, garden trees sense that buildup of tanninlike bitterness, and begin to pump . . .

"Milk," he decides.
And adds a drop from the antique porcelain jug on his tray.

. . . a milklike fluid down. That joins with a tremendous number of the bitter tannin molecules, so that what's left . . .

"Better," he decides, taking a subsequent sip.

. . . is sweeter, and won't hurt the tree.

"Hi!" his wife calls over, emerging from the back door with the fresh herbs.
"Hi—Hunnnh," he finds himself saying, in distress.
Feeling a marriage-threatening, dreaded pollen sneeze coming on again.

The problem is not that he *wants* to sneeze, or even that there's an especially high level of micro-dirigible pollen spheres bobbing in the air today. Rather the car exhaust fumes from the nearest roads have floated over, and reached his steadily inhaling

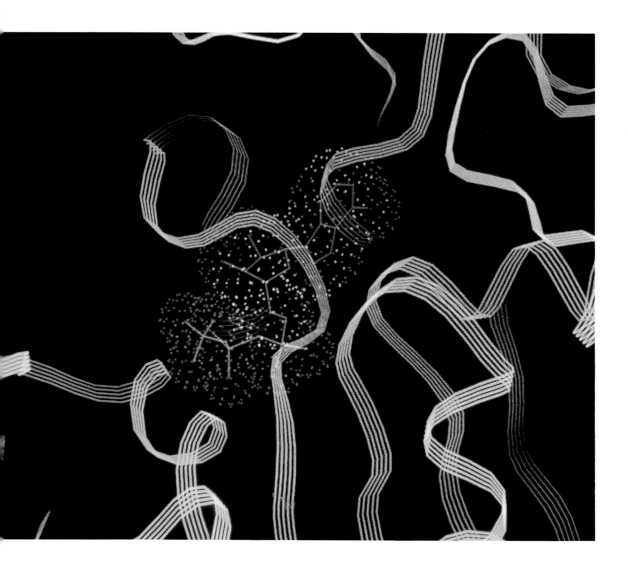

Identical in both plants and humans, this ATP molecule carries fresh energy to and from stored glucose. The green ribbon, rich in carbon, attaches to the glucose.

THE SECRET GARDEN

nostrils, and knocked out the tiny hairs on the inside of his nose which normally filter the worst of the pollen spheres away. That's why you're more likely to sneeze from hay fever on the edge of a city than far out in the middle of the countryside.

But . . .

"Uuuuunnngh," he chokes back.
The oak's leaf islands swaying above.

. . . remember those tree's roots. Not only can they vacuum in the nutrients they need from the soil cities around them, but they can also dump out into the soil other chemicals that they don't need. Some of these will simply be the excess from the incoming poison gases that the other garden plants are using against them. But some of them will also be the car exhaust fumes—especially the sneeze-enhancing nitrogen dioxide. The oak collects these molecules on its huge number of leaves and dumps them into the roots, still boring away under the reclining human. Even a smallish oak might have forty thousand leaves. A lot of the gunk floating past a tree gets caught in the air caverns inside each leaf; even more gets collected in the miniature pumps connecting those caverns with the tiny solar furnaces on each leaf. With a size of perhaps 6″ × 3″ for each leaf, there's twelve thousand square feet of leaf steadily pump-filtering up there. Along with the grass blades on the lawn and the other plants, that's enough to . . .

"Uhhnnn . . . uh," he finally says.
Triumphantly not sneezing.

. . . soak up, and ultimately pump down for safe disposal in the soil, dozens of pounds of pollutants each year—including those nostril-itching nitrogen dioxides.

"What?" she asks, wondering why he was making those faces.
"I . . . I was wondering how the cucumber's doing?" he tries.
"Oh," she says, and begins to Tread across their land.

We'll ignore, as usual, the devastation to the garden soil: the collapsed soil tunnels from that mysterious moving Great Weight above; the desperately fleeing computer

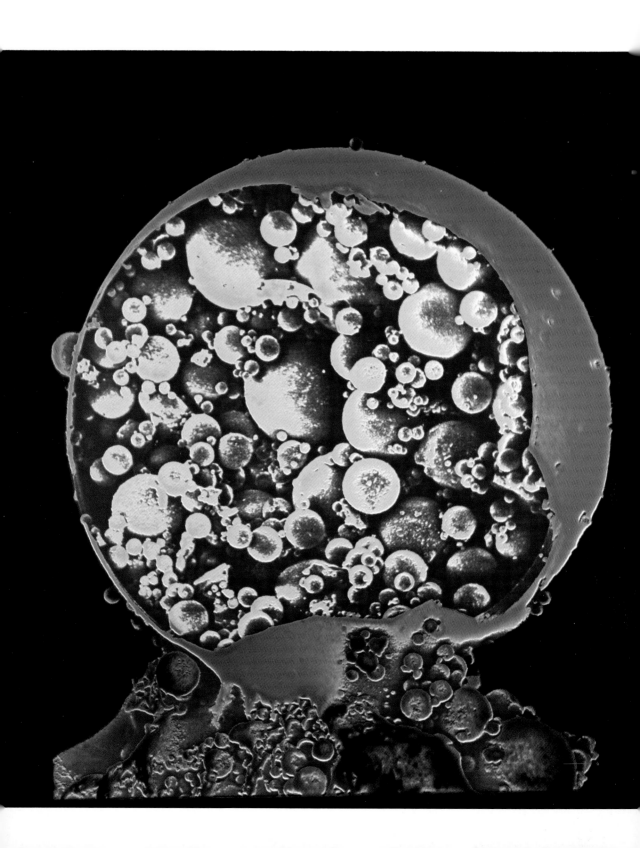

Worlds within worlds. Fly ash is a common microscopic air pollutant. When the outer sphere lands, it bursts open, releasing the smaller ones. What lands on a healthy tree leaf gets broken down and decontaminated, in large part; what lands in human lungs stays there. Magnification 10,000 times.

ants and the rest. We'll even ignore the Finger Crushing of the new geranium's surface, as the woman thinks she's gently rubbing her fingertips across it, but is thereby splatting into uselessness the defense maze chemicals which the flower has painfully pumped up in the lunchtime interval to defend against the scurrying mite-cars. And we'll ignore the playful Kicking Up of soil surface fungus capsules as she walks, which land on the cucumber and force that struggling plant desperately to spray out more ethylene vapor to defend itself. But what we can't ignore . . .

"You do not belong," she murmurs, leaning forward, as she picks the brightly colored caterpillar off that cucumber.
And flicks it away, unseen, to land deeper into the vegetable patch, on the tomato.

. . . is the caterpillar which has just been flung down onto the silent tomato plant. The caterpillar is going to orient itself after its sudden flight, grab onto the tomato plant, stick some nice holding claws in, and then settle down for a bout of hydraulic jaw—powered dining. It seems unfair for the caterpillar to take advantage of such a simple, drowsing, vulnerable plant as the tomato. But it's not unfair:

Not when the tomato gears up to full power.

First, only dimly sensing, the tomato locates the caterpillar-leaf, using the oak's trick of detecting debris from the incision point of the bite. Only then, with the intruder's location identified, can it switch on powerful defensive engines inside, and send something back up *against* that marauding attacker. But which chemicals could that be? Really strong ones, such as the tannins which the oak made, would certainly destroy the tiny caterpillar, but they'd also destroy the relatively tiny tomato plant too. And weak ones would just be swallowed by the caterpillar with no problem.

Caterpillars hatching. These mobile eating machines break out from their eggs head first and almost immediately begin feeding, first stripping just the waxy surface from the leaves; then, when

THE SECRET GARDEN

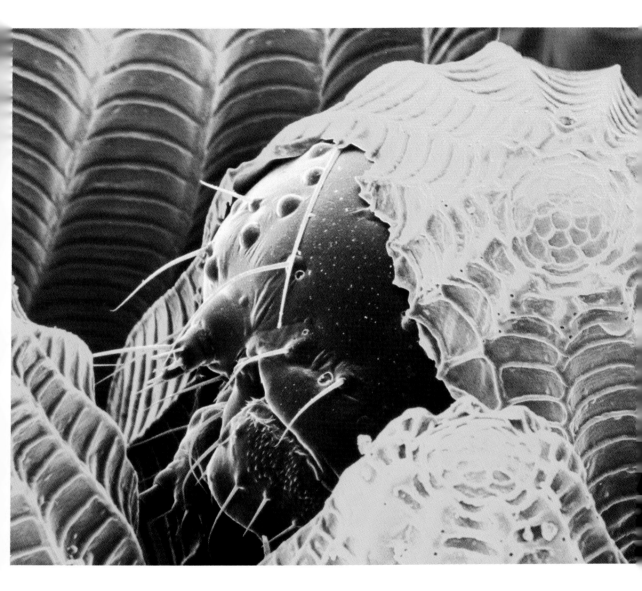

they're stronger, chewing through the whole leaf. Note the multiple legs helping them step over the micro-spikes and other obstacles on their leaf homes.

THE SECRET GARDEN

The chosen chemical has to be something which caterpillars depend on, which caterpillars absolutely need to live on, but which tomatoes aren't affected by.

Is there one?

"Um . . . Jenny?" he calls over, inquisitive, wondering if he can wheedle her into going back to bring him some tasty high-protein snack, to replace that miserable cracker.

The muscles that keep his wife bent over checking the cucumber plant, and the husband holding his newspaper to the side and peering—and the caterpillar chewing, and the ladybird flying, and any distant cows grazing—are all built from raw protein. Tomatoes can't get at that directly, but consider how we control our muscles. Without some triggering chemical for our muscle protein, our newspaper would seem impossibly heavy, flying creatures would lose power, and there'd be a steady splat of downed ladybirds, wasps, and aphids around the now-slumped human.

Luckily, we all have muscle-starting chemicals suffusing our arms and legs. But of course if those had nothing to oppose them the effect could be *too* strong. What *are* you doing up in that tree? your neighbor asks distractedly, as she brings back the lawn mower. *Aw-waa-ooo-wa-gaaaaa!!* you answer her, crashing two fists Tarzan-style on your now surprisingly muscular chest. What we also need is a chemical to reverse the first one. Not too much, for then we'd be in for a sudden crash-landing again, but just a little.

Such a second controlling chemical exists, and all through a quiet afternoon we're pumping both sorts around inside our muscles: automatically squirting first one, and then the other, so the levels don't become unbalanced. Clambering down the tree, the ant is pumping tiny, carefully balanced, amounts of these two chemicals. So is the powerful wasp hunt-flying overhead. So is the caterpillar as it meticulously holds the new leaf to eat.

And so is the tomato plant, as it now gets ready to grab inside this caterpillar and fling it off.

It seems like something from science fiction. It was just barely believable that a leafy shrub could produce an SOS chemical to call in a wasp. And that the tree, giant-size and made of complex woods, could use some of its complex structure to pump fluids around and protect itself. But not this. Tomatoes are supposed to be just dumb reddish-colored spheres. We, and other garden animals, are supposed to be the more intelligent, independent ones. We're not supposed to need to worry about something reaching up from the apparently safe gentle chemical *plant* side of the great divide and poking into the internal muscles that only animal creatures should have, and then, once it has a good grip there, *yanking* us back down.

But the tomatoes do just that.

No one knows exactly how tomatoes first developed this ability to create exact copies of protein-controlling molecules, which only animals should naturally have. Yet somehow—and starting just a few minutes after an insect bites into a leaf—a tomato plant begins producing this defensive protein controller. It pumps the stuff back to the attacked areas, and if several leaves are attacked at once, it even makes the intelligent choice of pumping the stuff into the youngest and most vulnerable leaves first.

Not only does your tomato reach into its attackers' proteins, but it also *learns* from the experience. The first time a tomato leaf is bitten into by a caterpillar, there will be a long ponderous wait, as the tomato works to get its response up to full power. But the *second* time the plant is attacked, it will have the DNA construction codes for the protein stopper under better control, and makes the stuff in what's been measured as about one-third of the initial time.

The defenses don't always work perfectly. There's a type of caterpillar that has outmaneuvered the sticky microspikes on some tomatoes' leaves, and uses them to narcotize itself so that it falls into a camouflaged sleep as the hunting wasps which are searching for that caterpillar skim over. But such design mistakes are rare, and with the protein-controller the tomato even has a fail-safe mechanism built in. To make sure that it doesn't suffer any self-damage as it makes the protein-controlling chemical, the tomato actually separates it into two different parts as it pumps the chemical up. Two different sorts of microcanisters in the tomato's leaves collect the separate parts of the chemical, and only one part gets nozzle-directed into each different canister.

It's a most elegant ingenuity. Either canister by itself is harmless, so if one leaks the tomato isn't damaged. But a caterpillar, even with tiny jaws, can't take bites which bring in just one storage canister. They're scattered too densely, and built too small, for that. This careful caterpillar, which seemed to be doing so well on this new green planet, gets thousands of these microcanisters spurting open with each bite. Soon its stomach proteins, leg proteins, and even the crucial hydraulically-controlled jaw proteins don't function.

The caterpillar might frantically try to make more of its own protein-starting chemicals to make up for the suddenly pierced gush from the canisters. But the only food source it has to draw on is these leaves which it's now painfully picking at. The individ-

Pond skating beetles are small enough that they don't burst through the taut "skin" of surface tension on water and so can run on the surface. The long legs add extra leverage for racing speed.

ual tomato leaves might be fairly small, but the plant as a whole, due to the wonderful hidden spaces under your soil, can draw food from root probes stretching for miles and miles in the clay passageways of those hidden soil cities. If your vegetable patch is well watered, and the plants aren't too densely spaced, then a tomato might turn its leaves into 10 percent defensive chemical by dry weight. To the caterpillar, those canisters have what seems to be torrents of the protein-seizing chemical.

THE SECRET GARDEN

As the tomato-leaf canisters keep on bursting, and the caterpillar finds its muscles increasingly switching off, it loses its footing, and falls off the plant. Just as with the oak, tiny tendrils begin to grow up from the tomato's topmost roots, pre-positioning so they can pump in the fresh caterpillar meat after it's been broken up a little by the soil bacteria. When you eat a tomato, there are quite likely to be molecularly recombined parts of such crawling caterpillars in there.

This muscle-grabbing by silently waiting tomato plants seems odd, but that's just because we're so much larger than the smallest garden beasts that we hardly ever have to suffer what they do. But . . .

"I'll try one," she thinks, leaning closer.

. . . bite into an underripe tomato. There's lots of the protein-controlling chemicals in the round fruit then. When they reach your stomach, or even . . .

"Bleech!" she says, spitting it out.

. . . your taste buds, you'll know how the caterpillars feel.

The trick of storing dangerous chemicals in two separate types of canisters is so good that . . .

"Something fresh," she thinks, hurriedly.

. . . certain other small garden plants use it. Mint plants have one microcanister in their leaves which fills with a potential nerve controller, and another which fills with a fluid that activates that controller. When they . . .

She crushes a fresh-cut mint leaf between her fingers, bursting the plant's microcanisters . . .

. . . mix, we get the distinctive smell which we identify with mint. It acts as a nerve toxin against insects—crushed mint is especially good against slugs because it blocks their usual neurotransmitters. In the case of larger humans, the mint simply switches up the specialized receptors on our . . .

. . . before, in relief, bringing it to her tongue.

. . . tongue, which normally work to detect cold temperature. With those receptors now firing on at much warmer temperatures than they normally would, anything you taste after eating some crushed mint will feel cool and refreshing.

"What's that?" he calls over again, concerned.

"Oh, nothing," she sing-songs quickly, to preserve her dignity.

Turning to go, she doesn't notice that Big Human Feet are now crushing numerous surface-resting vegetable leaves in her rush.

The hurried escape from the vegetable patch doesn't affect the crushed tomato leaves too badly: the instructions for defensive canister fillings had already been sent to them as a result of the caterpillar's precision biting. The problem is for the other vegetables. How can they tell the difference between a one-off crushing Foot and the beginning of a steady jaw-crushing caterpillar attack? Potato leaves start signaling a dangerous attack, as do pea leaves, and any other plants our Escaping Human has stepped on. Ponderous gushes of defensive poison goo slowly begin to be pumped up into the different plants, and even some of the roots get carried away, speeding the rate of production of their subterranean gas defenses. Traces of their fighting vapors rise up through the smallest microramps in the soil behind the fleeing human female, but the quantities are small, and those aren't going to bother her.

It's the exploding hydrogen bomb directly overhead that's going to bother her.

"You've got it?" he asks, looking up to his wife, finally standing near in the glaring sun.

Admittedly it's pretty far overhead. But the sunlight—which has escaped from the live thermonuclear furnace which is our sun—still enters the upper atmosphere at a little over 660,000,000 miles per hour. Some of it is absorbed by the now-famous ozone layer up there, but the rest continues on down, to whack into any exposed human skin. Suntan lotion does help a little bit, but we're also aided by the small floating islands under our skin, beside which are even smaller reserves of liquid chemical that activate as the light comes in, squirting on the islands and making them darker and darker, producing suntan or freckles in people with white skin, and little or no change in people with darker skin. But it's really wiser to . . .

"Yep," she says, stepping closer, under the tree for its shade.

. . . let some other thing take up the sun-blocking position. Yet how do trees, and grass blades, and the other unshaded plants survive their hundreds or thousands of hours of direct unblocked sun each summer? In part it's due to their thick waxy coatings and shiny light-reflecting covers, but even those get only part of the incoming blasts. Almost all sun-toasted plants also have a tiny robot chemical, constantly skimming alongside the DNA in their cells. If the long controlling strand of DNA underneath it is in good shape, and not too sun-damaged, the robot doesn't slow, and keeps on skimming. But if there's a problem, if one of the small connecting rods (a pyrimidine diner) in the DNA is cracked or splintered so that it sticks into one of its neighbor rods, then the robot saws through the DNA where it first detects the damage, and skims along until it reaches the bottom of the damage, where it saws into the DNA again. Then it simply continues on its way. The sawn-off fragment breaks from the main long strand of DNA and floats loose, like a single thin wood disk rolling off a log in a sawmill.

The process has to be quick, for otherwise the sun-damaged DNA part will start to copy itself. But the skimming system is an ancient one, and the chemical robots—in your oaks and roses, in the grass blades and geraniums, and in nearly every other green thing—almost always get there in time.

What is curious, and most satisfying, is that *we* have these little skimming robots too. There are vast numbers in our arms, and fingers, and everywhere else our body has exposed skin. We even have an additional, second robot that only activates the night *after* we've been exposed to a strong dose of sun. This second robot chemical attaches itself, suction cup–like, to the most delicate parts of our damaged DNA control strands. It can't do the welding repair itself, because it needs a precise burst of blasting energy for that. But all it needs to do is wait for the next day, when we emerge from our house, feeling a bit sore perhaps from all the sunlight the day before, but yet—with typical human repudiation of all previous day's vows—going into the sun again.

That nice, energy-blasting sun.

This is the needed energy source. The sunlight explodes the suction cup–like re-pairer, and instantly melt-welds the once-damaged DNA strand back into place. The DNA repairer is a solar-powered microbattery, conveniently in operation under our skin.

Before these processes developed in terrestrial biology, all life forms had to exist entirely submerged in several yards of ocean water, to be safe from our awful DNA-mutating sun. Even today, the first sort of robot skimmer starts working in your skin the moment you step into direct sunlight. By the time you've walked across your lawn and settled under a tree, the sun-shattered DNA it slices loose will have entered your bloodstream and, in de-assembled form, will be ready with the moisture in your breath to . . .

"It'll really help?" he whispers, a little ashamed.

Removing the apple he's vainly started eating to reduce the leftover lunchtime tastes, and so letting moisture-laden air gust from his mouth, heading her way.

. . . gust from your mouth, along with the remnants of all the other strong chemicals you might have ingested today.

"Yes," she says, as the garlic blast still there registers.

Reaching into her herb package for the antidote.

Some plants have managed to turn this sunlight-defense system into yet another weapon against ingenious caterpillars. There's one herb growing in your window box that usually has fewer insects on it than the other ones. That's because it produces a chemical much like the one that our own body produces for sunlight repair. That chemical is cheap to make—it doesn't need any inner energy source, remember—so the small plant can pour it out in quantity, even without anything like the oak's giant system of accumulating roots.

When a sprig of this ingenious . . .

"But what is it?" he asks.

"This," she says simply, handing him the fresh sprig of parsley.

. . . parsley is swallowed by an insect nibbling in some extra-early feeding, in the earliest morning when everything's dim and shady and the sun's not yet up, there's no immediate effect. Perhaps the insect even moves away, to feed somewhere else before coming back. But remember the solar-powered battery. Wherever the insect has gone, however far away, as soon as the sun comes up, the chemical is activated, and explodes in a burst alongside the DNA it's clinging to. And while our repair chemical fixes connecting rods when it's energized, this parsley-made one breaks them.

Sometimes you'll see an insect surviving on parsley (or the closely related parsnip) during the daylight. But in order to do that the insect will most likely roll the plant's leaves over itself, to form a green, shady—and sun power–blocking—personal umbrella.

"Hey, it works," he says, chewing, feeling the strange parsley chemicals cool the leftover garlic.
"Let's see," she murmurs, and promptly leans forward to kiss him.
At which time, in pleasant astonishment one apple core gets jettisoned into the bushes.

Readers of a romantic nature should perhaps skip the following sections. For the rising vapors from even a partially chewed apple core are likely to attract, on this summer's afternoon, a mobile cloud of tiny fruit flies.

And these creatures threaten, radically, our traditional notions of love.

The tiny soaring flies are attracted by the alcohol vapors from the apple. But although mobile and energetic, they're also almost blind, with a visual resolution less than $\frac{1}{100}$ of ours. They can't even see the Fleshy Bipeds, who now—to the continued delighted surprise of one of them—are gently kissing in a shared kneeling stance under the tree. Instead, the skinny microflies are concentrating, very hard, on simply trying to identify each other.

It's an important task, because the creatures only have a few days of life left. Somewhere in that whole descending cloud there's the best mate for each one, and what they need to do now is find each other. The problem, though, is that as they are nearly blind, it's hard for them to identify this best partner. A female shouldn't just settle for the first male to land behind her and sidle up, nervously ready to begin the gymnastically intricate mating task. He might, after all, be a pathetic, unmuscular, and most-unsuitable-for-parenthood specimen. Perhaps in other species the female could . . .

"Mmmm," she says, peeking open one eye to see his muscular arm on her shoulder.

. . . peek open an eye to look, and confirm that her potential inseminator was indeed muscular, virile, and so forth. But not here. The female fruit fly needs some nonvisual way to confirm that he really is strong and suitable.

There is a way that she can solve this problem. Male flies, if left alone, naturally produce different sorts of odors. Strong, wing-toughened, he-man microflies produce a strong, toughened, he-man sort of odor. It comes directly from all the glisteny muscles around their wings, and so it should be impossible to fake. Skinnier fruit fly males have

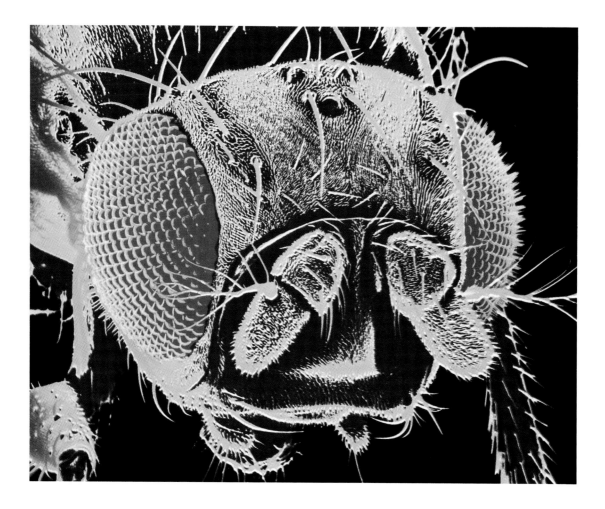

A friendly fruit fly. Despite its apparently large eyes, this pleasant creature is so small that it registers little light and lives in a perpetual dim haze. Mating is a treacherous process, where each sex produces vapor clouds to mislead intended partners about prowess and size.

no choice but to let off a correspondingly weaker odor. The female merely needs to sniff what's seeping up around her, and from that she'll know if there's a tough prospective mate sidling up.

But can she really trust what she sniffs, there around the jettisoned apple? What if the male is actually a weak, scrawny, and wimpish specimen, who has found some way to biosynthesize those molecules which are supposed to signal strength? Then she'd be stuck. It's not impossible, for the chemistry is only a little different from the usual chemistry a male fly undertakes as he hurries through your garden air, looking for some nice rotting fruit to provide the desired home for his new mate, and the satisfactory completion of their life. It would be no more unscrupulous than for a balding accountant to rent a toupee and a Porsche to impress his date. Who can trust a date-desperate Man?

A large proportion of the scrawniest male fruit flies do prepare these false-identity chemicals. They pump them into little reservoirs under their wings, and then when they're close behind a landed female, they fan their wings—we hear it as the distinctive 250 Hz low-whine of these flies—to push some of that potency-suggesting chemical forward. It seeps around the female's feet, and floats up, enveloping her body, and soon seems to make her lose all control and fall back against his straining skinny body, totally taken in by the subterfuge.

Or does it?

"I wonder what she's thinking," he thinks, toes wriggling, as the delicious afternoon kiss goes on.

"This'll get him to mow the lawn," she thinks, pressing her lips closer to guarantee it.

The male fruit fly won't bother to waste his deception chemicals on an already mated female, so before he started the whole wing-pumping subterfuge, he wriggled his antennae around a little to confirm that she was exuding the odor that signals an unmated, *virgin* female. His own sperm supplies are limited—as is his very lifespan—and it wouldn't do to waste either of those on a female who's already mated. Yet this virgin chemical has been floating back from the female since the male landed behind her, so his brain circuits are satisfied that he's in the right place.

Which is pretty stupid.

The female has been false-duplicating this chemical all along. She's mated already, and the genetic material for her children is already developing within her. If she let the present male know that, though, then he'd step smartly back, flap his scrawny wings hard, and take off to find another, more suitable, *true* virgin female somewhere in the traveling cloud to mate with. And what would that do? The result would be a set of infant flies, someday soon, which would be let loose in this garden and would then be around to compete with the first female's own children. That's not good for her own children. Who wants that competition? Ivana might not have had a deep and profound love for her billionaire, Donald Trump—she might even have been the first to guess that he wasn't a billionaire at all, but was just renting his toys from the banks—but in the early days she still didn't want him to sire children who would later compete with her own. If that meant persuading him to think she was still desirable, still available and enamored, then she'd do it.

For the female fruit fly, the strategy is the same. And just as Ivana got the money for all her jewels from Donald, the cunning female fruit fly gets the chemicals for her own virgin-suggesting charade from a raw chemical that she's been extracting from the now actively mating male. She soaks it up, turns it into the false-virgin scent, and—with her own tiddly wings flapping in apparent delight—coolly sends it back in his direction. The poor male, still actively mating away, is hoodwinked, fooled, conned.

Or is he?

"Mmmm," she murmurs, his wife, near-smiling, as the luscious kiss continues. Thinking of the nice dry lawn which she will soon watch being mown, before their evening guests come, without any labor from her.

"Mmmm," he murmurs, her husband, smiling, thinking of how he's set the timers on the new sprinklers to come on any moment now, so that he'll escape any of that mowing.

A third level, and then our crosses and double crosses are done. Did the male fruit fly *really* transfer that raw chemical to the female just by chance? Remember that the flies live in a different perceptual world from us, and even on a sunny afternoon have to make do almost entirely with what their sense of smell signals to them. The first male "knows" that he's not going to be the only one who'll be attracted to any fake-virgin smell he can get his partner to send floating up. It'll spread across a distance of millimeters, and sometimes even of mighty centimeters, to bring in *other* male flies from the cloud. So long as they crowd around, chemically believing that there's a solitary female down there for them, they're certainly not on the hunt, elsewhere in the cloud, for any *true* virgin females. The first male, that is, has made sure that although he might be fooled and kept away from true virgins . . . the others are going to be fooled, too, improving his odds of mating with the real thing.

If humans were as coolly efficient as fruit flies . . .

"Hey!"she says, standing quickly, as the cool sprinkler water suddenly mists down.

Neither partner notices her feet digging into the lawn edge.

Or the sudden flashes of sunlight, which have triggered the long-dormant crabgrass there to emerge.

. . . there might well now be forceful discussions of who should have applied the herbicide to avoid such weeds appearing, and who must now get down and finish the weeding, before the guests arrive. But thanks to mere human beings . . .

"It's raining," she says.

"I know," he smiles, taking her hand.

"Oh," she says, but accepting his touch, and heading back to their house, where some more long chemically restoring rest is in store.

. . . the garden will, one final time, be left to take care of an intruder on its own.

The problem began long before this season's gardening. Weeds live an extremely long time in the waiting seed stage. Five years is common, a dozen years is not unheard of, and there are even reports of a seed which has been caught in Arctic gravel and revived, to grow successfully, after a dormancy stretching from 6,000 B.C.

Weeds protect themselves well during this long wait. A common crabgrass seed will spray out tiny amounts of an antibacterial poison as it waits, so that no soil bacteria can come close and disturb it. Think of it as a meteorite that has crash-landed in one of the microsoil cities, blocking off some passages entirely, and sending strange throbbing steam vapors down many others. Once it has puffed out enough poisons to clear a safe space, the seed will even, in a spirit of maximum ecological efficiency, swallow back any excess of that expelled poison it can. That way it won't run out of its defending poisons as the waiting months and years pass.

Because weeds can survive so long, an ordinary garden is likely to contain a residue of almost every individual weed plant which grew there in the past years or even decades. And because the seeds are so small, and the microcities down there have so much stacked-tight floor space, the total number of weed seeds that might be lurking is quite surprising. The first times counts of these seeds were done with modern equipment the botanists went back and recounted, because they couldn't believe they were getting the right figures. But the repeat experiments came up with the same figures:

there can be up to thirty *thousand* separate poison-spraying weed seeds waiting under a single square yard of lawn. It would be unlikely if the figure for an ordinary suburban lawn is much less than a thousand or two in each square yard.

Normally the weed seeds stay below, locked away in the cold and dark until they run out of supplies, and the clay city bacteria finally get close enough to destroy them. But heavy digging—such as might occur when you prepare a hole to plant a new flower or even just when you gouge a shoe or a toe deeply into an exposed spot—will allow in that single intense flash of sunlight which signals that the surface is now accessible, and to which the weed may respond.

And begin to grow.

Little root arms grow out of it, of course, and the levels of expelled poison gas increase. But the main activity is its vertically rising digging probe. If the weed has been waiting an especially long time, its miniature battery packs will have been drained so low that it has to get this stage over with very quickly, or else it will run out of energy. To do this it boosts itself by puffing out ethylene gas of its own whenever it reaches a thick portion of soil. That makes the roused weed machine strong enough to push through.

Once it reaches the surface, the weed unfurls an antenna, a structure of tiny leaves, which turns slowly, taking coordinate readings of the sun's location at this strange above-ground site it has arrived in. Then it unfurls its larger solar-panel leaves and aims them in the right direction, to send back energy from the solar furnace in the sky. The transformed chemical energy is pumped back down so the ancient battery pack, mission achieved, can begin to switch off. This sunlight-derived energy also powers more root-probes, down below, and sends them stretching into the mineral stores around the grass roots.

The weed machine now grows far more quickly as its new roots begin to spread. It increases the level of destroying poisons it is spraying out, and also begins preparing

Not big oak roots, but the microscopically thin living fungal arms (magnified over 500 times), through which nutrients can get passed between the roots of a lawn. A miniature microphone attached would reveal the gurgling of the fluids pumping through.

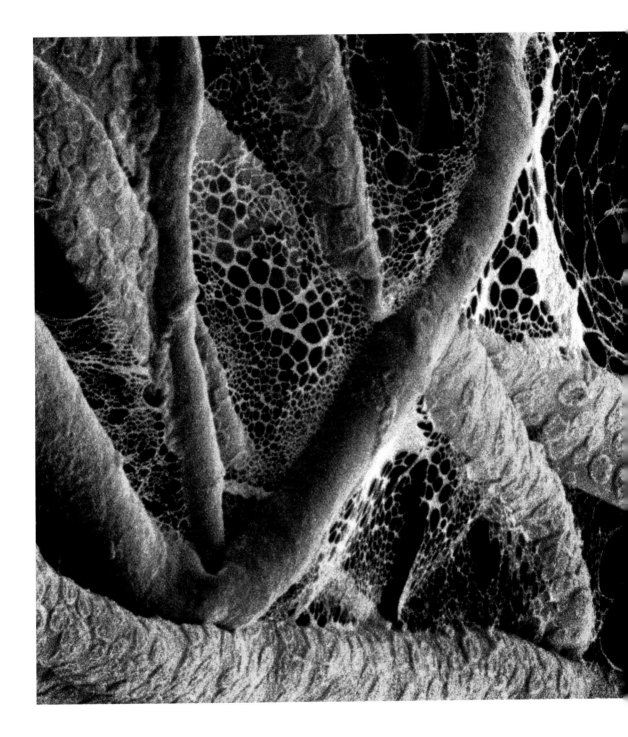

seeds to create the next generation of weeds. At this point the weed is vulnerable to a contact herbicide such as 2,4,D-T. That herbicide contains the chemical auxin, which, once absorbed by the unfurled top leaves, begins to force the weed's internal machinery to work too quickly. But if herbicide isn't applied, the plants adjacent to the weed will have to get rid of it themselves. Of course this doesn't work perfectly, because weeds are likely to appear for a while in almost all lawns. But, just as with the fungus growths, each time our fields and lawns are not entirely taken over by weeds, we're seeing these self-protective measures successfully at work.

To some extent the ranks of lawn blades try to fight back by sending countering poison clouds through the soil city tunnels, to strike against the new weed's roots. Sometimes this works well: certain grasses can create dilute bursts containing hydrogen cyanide gas. But usually the weed is stronger and, as the cucumber did, it opens root hatches to absorb the incoming poison gas, and transform it into something stronger that it can send back out.

What the threatened lawn must do is take advantage of what makes it different from the weed. For where weeds are, par excellence, solitary hunters, lawn grass is overwhelmingly a tightly grouped species. Most grass blades on your lawn are over eight months old, and some will be over twenty months old. But even more importantly, the whole community of lawn grass might have been growing in the same place for years. And in that time, all stable and steady, it will have had time to acquire friends.

Lots and lots of miniature friends.

In the late afternoon now, through the hours till the couple return for their final pre-dusk drink on the patio, the residents of our original underlying soil cities will be hurrying forward to help. As the first grass blade is attacked, it sends out chemical sprays from its roots—a mix of vitamins, amino acids, enzymes, and sometimes particular hormones—which alert the soil city creatures that something unusual is going on in that vicinity. It's not just the gliding bacteria which are quick to speed over, as they did with the new geranium. In the case of a long-established lawn, there will have been time for another, even more important, creature to come.

This is the E.T.-like fungus, once again. It winds one living arm through the soil city tunnels until it reaches the grass blade's root. Sometimes that living arm doesn't survive very long and is just offered up as a living roadway for the gliding bacteria to speed along, taking replenishing nutrient "bites" from the fungus arm as they go. But the fungus creature is not limited to a single arm. If conditions are right, the fungus can grow many hundreds or more. Those are what reach to the attacked grass root . . . and then there's a pleasant surprise.

Apply a tiny microphone to the particular fungus arm that's reaching to a grass root under attack from a new weed. From inside the leathery arm comes a gurgling, as if something were being pumped along inside. It's fresh supplies of luscious, replenishing sugars and other foods. And the fungus isn't pulling in fluids just from the few nearby clay cities. Those, alone, could quickly run out. The fungus has also sent one of its *other* arms in the opposite direction, to another—healthy and quite unattacked—grass root, and it's siphoning over nutrients from there.

This is help the solitary crabgrass weed could never get. There's an entire cabled-up network under your lawn, connecting all the grass roots. It's been building up for the whole life of your lawn and by now can easily shunt liquid food supplies from a well-supplied sector to a threatened one. The weed, though tracking the sun with its antenna and trying like mad to grow fresh roots of its own, can't compete with a system this large and long-established.

The size of this automatic cabling system is difficult to comprehend. A single square yard of lawn planted with Kentucky bluegrass can have ten billion root probes. Even if only a fraction of them have had time to build up the connecting fungus cables, that's still many millions of links, and so an extraordinary volume of subterranean space from which any threatened grass blade can, via the shunting network, draw help.

Lawn grass isn't the only plant in our garden which is cabled together this way. Roses, and vegetables, and even, if everything finally works out, the geranium will to varying extents be hooked in to the fungus-linked system. In some places the sub-soil links will stretch just a fraction of an inch; in others they might go on for yards. It's all invisible to the naked eye, and usually kept entirely in the hidden world below, untouched by any human hand.

Though not always.

EPILOGUE
DUSK

"Hey!" he says, suddenly disturbed from his quiet reading on the patio at this hour of dusk now, by something very small flying about him.

"Hmm?" she asks, calmly, readjusting in her chair to get better light.

"I think they're back," he whispers, grabbing for the newspaper to roll it up.

"It's not a bee," she says, still not looking up from her paperback.

"How do you know it's not a bee?" he asks, striking at the air with the rolled newspaper weapon held staunchly before him.

Trees are hooked to and partly fed by the slowly gurgling cables as much as other plants. So squeezing a rolled-up newspaper means, most pleasingly, that you're just squeezing molecules which were until recently streaming through such subterranean fungus cables.

"Because it's a moth," she says, finally looking up, and pointing to the back-door light. "It's getting dark now, and they like the light."

"Oh," he says, chagrined.

"Did you hurt it, Mike?"

"Me?" he asks, surreptitiously wiping away the dust which appeared on his newspaper where he whacked at the disgusting thing. "Hey, I like moths."

"But where did it go?" she asks, trying to gaze into the now lowering gloom over their garden, where somehow this creature has vanished.

Tumbling low, disoriented by the attack from that maniac with the newspaper, the moth might head straight into one of the sticky spider webs, drawn taut in the shrub beside the trees.

But this is no problem to a night-flying specialist. It just slips off a few more of the slippery scales from its wings—this was the source of the "dust" on the newspaper—and so disengages. The spider might hurry out to try to get it, but at this hour well into dusk the air temperature has become too low for the spider to wait, as it had through the day, in the exposed position near the center of its web. Instead it now has to wait in a more protected position on one of the nearest shrub leaves. By the time the spider reaches the center of the web, this moth is gone, flying off again.

What seems to be a more serious problem, as it tries to orient in the dim air, is the waiting . . .

"I'm sure it's okay," he says, also trying to peer into the garden, where the day's fading light is so low that it's hard to see.

"Oh?"

"Why it's nice out there. Just look at the nice bird."

"Where?"

"There," he says, pointing to a small outline gracefully arcing through the air just yards away beside their tree.

. . . killer bat, hurtling into this garden now. Bats are far better fliers than birds at night, because they don't have to rely on vision, but can navigate their way to any waiting insect food by sending out sonar pulses. The echoes that bounce back give them a picture of what's moving in the air in front of them—it's like the oscilloscope vision an air traffic controller builds up on his airport radar.

The moth's first defense again comes from those fuzzy scales it has all over its body. To us they just seem ungainly, a mistake. But because of their uneven shape, they give the bat only a fuzzy outline on its sonar scope. This alone wouldn't be any great problem for the bats. They immediately switch on a higher searching bandwidth—like a radar operator moving to an emergency frequency, if they're having a problem getting a clear image at the first one.

Some moths can't respond to that, and for them that first sonar contact by the bat is likely to be followed by a quick ingestion, and an almost as rapid transformation into fuel to help the bats hunt more moths. But our errant moth is not going to be so unlucky.

As the bat gets that revised sonar fix and swoops in closer, the moth can hear it coming. The moth has a sort of sensitive microphone built out of a tiny rigid plate below its wings. This microphone device locates the swooping killer in the air, even in dim, dense underbrush or where there's no light at all. The moth doesn't have to be able to visually "see" anything at all to get an exact fix on where its attacker is. With the new information, it just changes course in the air, and flies away.

For some bats that would be it, and they'd have to start again, searching for a new moth. But some garden types have an additional feature. These bats switch on yet another part of their sonar apparatus, which measures the degree to which the sonar picture they're getting of a target is *moving* as they locate it. Its results are the the same as the Doppler circuitry the most modern of human radars use—only the bats were hunting with it many millions of years before. With that degree of sophisticated sonar the bat can home in even on the moth's abruptly changed course. It accelerates a final time, spreads its devil wings wide . . .

"Such a sweet bird," she says, straining to see.

"So acrobatic," he concurs.

"Wow, that was fast," she says, seeing something else now.

"Whoa," he agrees, seeing it too.

"It was . . ."

". . . like a pirouette," he finishes for her, proudly.

. . . and misses. How could moths survive if they let bats nab them with a measly Doppler analysis? The moth, apparently stumbling so unevenly through the garden air, waits till the bat is just about to reach it, and then, neatly, blasts out an ultra–high-frequency jamming signal. That wipes out the bat's scanning picture—as if a radar screen had suddenly gone blank. To be sure that the bat won't get it once it's recovered, the moth is even likely to tuck its wings tight to its body, and go into a plummeting fast dive, before twisting sideways, and starting off again.

Having evaded all these obstacles, the moth can get on with what it initially set out to do, which was, simply, to find a mate. Somewhere in this garden, or the next one, or perhaps even the one after that, there's a female moth, clutching to a tree and waiting for this male suitor. She can't lure him by sound, as that would give away her location. Also sound isn't very accurate over these distances. She can't lure him by bright colors either—they would be of no use in dim twilight.

Instead . . .

"Mmmm," she says, leaning close, her perfume wafting up nicely to her mate.

. . . the female moth sends out odor molecules. There's nothing mischievous about them, as with the fairy flies. With the still air at dusk the invisible molecule trail can spread, expanding neatly, without getting too dispersed. The molecules form a wide tunnel in the air, which the male moth, now free to search for, finally locates. He enters it and heads along, carefully tracking from side to side every now and then to be sure that he's not left the directional cone they're marking. To us it might appear as errant bumbling, yards up in the "empty" air, but it's not.

Down below, on the lawn, the invisible trails of the ants are not doing as well. For . . .

"Chilly?" he asks.

"A little," she says, shuddering in the sudden cool of the dusk air.

. . . most of the ants, being cold-blooded and entirely without furry insulation, try to move back toward their nest. But their chemical walkways need to be steadily res-prayed, by ants lowering their nozzles and letting out the right vapor. Otherwise they break apart, and float away. The last ants left out begin to wander, disoriented and lost, as these crucial return walkways disappear.

When they finally do make it back they'll find that the other ants aren't even in the same places in the nest as they were before. With the growing cold the ants will have begun to walk deeper than usual in the nest, where more heat remains. There, all together, they start to huddle for warmth. The latecomers, unable to push into the huddling group in time, will likely have to stay on the outer edge of it all. They'll get colder, and their nerve circuitry will slow down, and they'll stiffen into the start of a night-long dormancy. Soon they won't even notice the last organizing ants, frantically hurrying down the underground ramps a last time, juveniles from the incubation chambers higher up by the patio tiles dangling in their jaws, bringing these future citizens to join the night-surviving huddle.

The bees do a similar thing. They've had to return to their hive because they too can't survive the night's drop in temperature out on their own—though bumblebees, with their hairy insulation, survive cool air better than most, and are likely to be some of the last bees you'll see out in the day. Also with the fading daylight they can't track their way out to the flowers. The visual image detected by the mighty hive brain gets dimmer and grainier.

As the last honeybees approach the hive, they'll be checked by the gatekeepers, and then let in for landing. Earlier in the day there would have been young bees clambering up on the walls of these entrance corridors, their wings fanning to provide living air conditioners. Now those younger bees have climbed back down and instead hurried to the queen, to keep her warm. The only times their huddle will break apart is when a team of house-clearance bees approaches, carrying something unwanted, such as a bee which has become too bedraggled from all the activity of the day and keeled over dead. Usually the clearance bees just dump the newly deceased right outside the hive, though sometimes they fly in elegant formation, holding it till they're several feet away, before letting go, turning around, and hurrying back in.

"Coming?" she asks, at the open door.

"Soon," he answers, looking up, trying to imagine again the last sun rays, up at the very top of the oak.

Although many of the aphids will be back at their usual small plants, a few are likely to be clustered up near the top of the oak now. Through the day they'd ranged on different leaves of the tree, as much as possible trying to pierce the nice juicy veins on the undersides of the leaves. But now as they too feel the increasing chill, some of them will walk or sometimes fly higher, to the upper side of the leaves or an exposed stem, where they have a chance of getting the last warming light of the day. The top of the oak canopy, perched way above the now dark garden, is the final place for that.

Once the last sun rays are gone, the aphids exposed up there will probably crowd a little closer together. Or they might find a convenient crevice in a leaf or twist of stem, and huddle there safe from the night wind, perhaps using their chemical communication signals to alert any stragglers.

In the tree itself, the great photosynthesis furnaces are now closed down for the night. The microscopic breathing holes in each leaf have begun to squeeze tight, and the tree is back to where it was before the dawn, puffing out carbon dioxide instead of oxygen.

"A final look?" he thinks, alone now.
Stepping out, one final time, across his property.

Down on the lawn surface, the day residents are entirely gone. The tiny jumping spiders, their clicking telephotos useless, are hunkering down for warmth beneath pebbles or any tumbled grass blades. They've been replaced by stealthy night-hunting beetles, sensitive to the infrared signals, or deep frequency vibrations, which are all that can be detected in the dark. Sometimes there will be downdrafts of cool air, bringing down from the sky flotillas of the tiny spiders we saw being launched earlier. This conveniently landed food will be part of the meal of these night hunters, but as the slowest of the night hunters will themselves be consumed by the earliest live spiders the next day, it's simply a convenient form of alternating food.

On the grass blades themselves, the climbing fungus is starting again, leaching LSD-like chemicals to keep any competitors at bay, just as it did last night. By dawn it's likely to have reached the tops of the grass blades, and started constructing living tubes from its own body, to feed off other grass blades nearby; we can sometimes see those nutrient-gurgling tubes very early in the morning, as fine white threads. A few pollen grains—their internal buoyancy bladders failing in the cool air—are likely to already be landing on this growing fungus, and its living threads.

A few of the drowsy water-bears which haven't been killed by residual pesticides may be roused from their sleep by the increasing moisture of the early night, and blearily latch on to any passing trouser cuffs, before returning to their deep sleep. Earthworms too will have emerged, attracted by the moisture seeping into their soil tunnels. And all around the grass blades, any of the errant E.T-like fungus capsules that have crash-landed earlier will be opening up, and extending their first tentative "arms," due to the same useful moisture.

"Hmmm," he says, kneeling at the lawn edge, one finger out, touching the new geranium, satisfied at its survival.

The sounds of car doors and the first arriving dinner guests echoing dimly from the far side of the house.

The geranium's roots are likely to have attached by now, with the poison droplet stumps on its leaves being refilled. It's also likely to be doing better in the under-soil battles with the vegetables, as all the stress in the vegetable patch earlier will have made those plants redirect energy from their roots up into their leaves.

Further down, deeper in the soil, the oak's great roots are also doing well. The amount of sugars which the small aphids extracted from the leaves was not enough to hurt. There remain great gallons of fresh sugars which those high-canopy leaf solar panels produced during the day, and they're only now reaching the farthest roots. Now when the leaves far on top close down for the night, these roots begin rumbling into full power, quickly reaching their top speed of several inches travel a week through the soil.

And even deeper, underneath them all?

"Mike?" she calls out, briefly worried now, not seeing him; at the back door again.
"Here, Jenny," he answers. Then he stands, brushing the last bits of soil off.
And walks back, content, to his wife, and to the light and their home.

Because the soil is such a good insulator, everything takes hours to travel into the depths. The night's coolness is still caught in the topmost layers. The late afternoon's warmth has reached a little lower, and only below that are the remnants of the noon-time direct heat.

Underneath it all are the first heated molecules from the dawn. Each time the heated molecules hit the level of a new soil city they wake it up. Now, at dusk, they've reached almost as far as they'll get before they dissipate completely.

In the deepest soil cities—the dark and pleasant corridors, the trundling soil mites and leathery fungus gardens—a new day has, only now, begun.

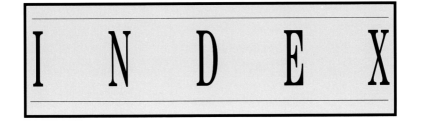

Page numbers in *italics* refer to illustrations.

sex chemicals of, duplicated by
    cucumber plants, 81
  triungulins, 100–103
  on willow trees, 112–14, 127–28
begonias, 27
birch trees, caterpillars on, 112, 124
birds, 56, 128, 129, 169
blood, human, *126*
  coagulants in, 108
braula, *87*
bread mold, *115*
breathing pores, leaf, *59,* 81, 110, 172
budding, tree, 132
bumblebees, 171
buoyancy chambers, of pollen, *119,* 172
butane molecule, *83*

cabbages, 60–61
carbon dioxide, 26, *59,* 68, 79, 172
car exhaust fumes, 141–43
caterpillars, 46, *146–47*
  hatching of, *146*
  leaf-repairing molecule in saliva of, 112,
    124
  of monarch butterfly, 135–36
  on parsley, 156–57
  predators of, 56, 57–61, 129
  on tomato plants, 145–53, 154
  on trees, 112, 128–29, 132–36, 137, 139
  webworm, 55–60, 133
chemical communication, 36–37
  aerial, 82, 119, 137
  by ants, 42
  by aphids, 31–32, *33,* 48, 52, 82, 101,
    119, 172
  among trees, 137

chemical defenses, 22, 36–37, 65, 69, 110–
    111
  of ants, 125
  of aphids, 31–32, 34
  of cabbages, 60–61
  of cucumbers, 78–81, 111, 114, 145, 164
  of geraniums, 70, 72, 73, 79, 84, 111,
    117, 145, 173
  of grass, 164–65
  of mint plants, 153
  odorous gases, leaf emission of, 58–63,
    101, 102, 114, 128, 133, 150
  of parsley, 156–57
  of pine trees, 61–63, 128
  of roots, 79–81, 110, 140–41, 143, 154,
    163–65
  of rosebushes, 50–51, 58–60, 65–68,
    101, 133, 136
  sunlight and, 156–57
  of tomato plants, 80, 145–53, 154
  of trees, 14, 61–63, 112–14, 127–29,
    133–37, 140
  of whiteflies, 84
  of willow-tree beetles, 112–14
chemical leaf markers, 83–84
chemicals, 14, 36, 64, 110–11, 172
  alcohols, 58–60, 102, 157
  anacardic acid, 72, 84, 117
  aspirin, 112–14, 127–28
  herbicide, 164
  hydrogen cyanide, 110, 164
  hydrogen peroxide, 87
  isoamyl acetate, 99–100, 114
  nitrogen dioxide, 143
  pesticide, 46, 48, 50
  protein-controlling, 150–53
  sex, *see* sex chemicals, insect

human beings (*cont.*)

    immune system of, *see* immune system, human

    muscles of, 150

    pesticides inhaled by, 46, 48

    saliva of, 108, 140

    skin of, effects of sunlight on, 154–156

    sneeze response of, 29, 119–20, 141–143

    tears of, antibiotic nature of, 123

    tongue of, 153–54

hydrogen cyanide, 110, 164

hydrogen peroxide, 87

immune system, human, *76*, 108, 111

    histamine produced by, 97–98, 120

infrared detection, *53*, 82, 119, 172

inorganic fertilizers, liquid, 76

insecticides, *see* pesticides

insects, 14, 23, 112

    mating behavior of, *see* mating behavior

    nitrogen in, 39–40, *71*

    plant control by, 31–34, 117, 124–25

    sex hormones of, duplication of, 135

    see also caterpillars; pesticides; *specific insects*

isoamyl acetate, 99–100, 114

ivy, poison, 72

Journal of Chemical Ecology, 14

Kentucky bluegrass, 165

lacewings, 48, 50, 51, *53*

ladybirds, 29–34, 41, 51–52, 56, 82–84, 113, 124–25

    leaves chemically marked by, 83–84

    pesticides vs., 48, 50

    sensory apparatus of, 30

lawn, *see* grass

leaf-contact pesticides, 45–47

leaves, 55–56, 65–68, 110–11, 112, 124, 154, 162

    alcohols in, 58–59, 102, 157

    breathing pores of, *59*, 81, 110, 172

    carbon dioxide breathed by, 26, 59, 172

    caterpillars on, *see* caterpillars

    chemically marked by ladybirds, 83–84

    circular rings on, 65

    exit tubes of, 135

    fungus on, 65–68, 69, 70

    geranium, 72, 111, 117, 145, 173

    liquid inorganic fertilizer sprayed on, 76

    odorous gases emitted by, 58–63, 101, 102, 114, 128, 133, 150

    oxygen emitted by, 26, 172

    repaired by caterpillar saliva, 112, 124

    tree, 112, 124–29, 132, 133–37, 143, *145*, 172, 173

    wax on, 57–58, 111, 155

malathion, 48

manna, 125

maple trees, 132, 137

mating behavior, 61, 81, 119

    of fruit flies, 157–60, *158*

    of moths, 170

    of tardigrades, 37–38

sunlight, 36, 154–57, 162, 163
   ant navigation and, 41
   chemical defense produced from, 156–57
   DNA damaged by, 155–56
sycamore tree, 124
*Symposia of the American Chemical
   Society,* 14
systemic pesticides, 50

tamarisk tree, 125
tannin, 140–41
   produced by rosebushes, 51, 133, 136
   produced by trees, 133–35, 137, 145
tardigrades (water-bears), 37–39, *38,* 42,
   46, 117, 173
   dormancy of, 37, 39
   excretory glands of, 37
   mating behavior of, 37–38
tears, human, 123
temperature, 39
   of air, 52–55, 61, 117–20, 123, 168,
   170–74
   human tongue receptors for, 153–54
   see also heat
tetanus bacteria, 116
thistles, 30, 46
threat displays, of bees, 93, 100
time-hiding defense, of trees, 128–32
tomato plants, 79
   chemical defenses of, 80, 145–53, 154
   ethylene gas emitted by, 80
   fungus on, 80
   protein-controlling chemical produced
   by, 150–53
   roots of, 80, 152, 153
tongue, human, 153–54

tools, *see* gardening tools
trees, 112–14, 124–37, 139–43, 150, 168
   aerial chemical communication among,
   137
   air pollutants absorbed by, 143
   ants on, 123, 124, 125, 129, 137
   aphids on, 124–27, 137, 172, 173
   balsam, 135
   birch, 112, 124
   budding of, 132
   caterpillars on, 112, 128–29, 132–36,
   137, 139
   chemical defenses of, 14, 61–63, 112–
   114, 127–29, 133–37, 140
   ethylene gas produced by, 136–37
   forced migration defense of, 128–29
   fungus on, 61, 124, 129
   insect-locating ability of, 125–27, 132–
   135, 136, 145
   insect sex hormones duplicated by, 135
   leaf defenses of, 125–27
   leaves of, 112, 124–29, 132, 133–37,
   143, *145,* 172, 173
   maple, 132, 137
   milklike fluid in, 141
   oak, 72, 127, 128–29, 132, 133–35, 136
   pear, *131*
   pine, *see* pine trees
   poplar, 137
   roots of, 112, 127, 136, 139–41, 143,
   153, 156
   sap of, 124–25, *126*
   sycamore, 124
   tamarisk, 125
   tannin produced by, 133–35, 137
   time-hiding defense of, 128–32
   willow, 112–14, 127–28, 137

triungulins, 100–103
Trump, Donald and Ivana, 160
2,4,D-T, 164

ultraviolet, in flower markings, 91

venom, bee, 96–98
    histamine produced by, 97–98
Vespula maculifrons (yellow jacket), 57–
    60, 102
visual sensitivity, 36, 93, 157, *158*
Vitamin C, *76*

wasps, 57–61, *94,* 121, 125, 128, 150,
    151
    sensory apparatus of, 57, 60
    Vespula maculifrons, 57–60, 102
    whitefly prey of, 60–61, 84
    willow-tree beetle prey of, 114
water, 37, *38,* 78, 81, 128, 140, 152, 173
    cold, as aphid spray, 32, 50
    correct application of, 70

evaporated from plants, 27, 111
    fungus helped by, 65, 70
    pond skating beetles on, *152*
    in soil, 26, 27
water-bears, *see* tardigrades
wax, on leaves, 57–58, 111, 155
webworm caterpillars, 55–60, 133
    eating habits of, 55–56
weeds, 80, 161–65
    leaves of, 162
    seeds of, 161–62, 164
whiteflies, 60–61
    chemical defenses of, 84
wild grass, pollen of, *131*
willow tree, 112–14, 127–28, 137
wing-beat sound frequency:
    of bees, 91, 93, 101
    of fruit flies, 159
worms:
    earthworms, 36–37, 46, 173
    nematode, 79

yellow jackets (*Vespula maculifrons*), 57–
    60, 102

Picture Credits